The Ocean World of Jacques Cousteau

Mammals in the Sea

The Ocean World of Jacques Cousteau

Volume 10
Mammals in the Sea

THE DANBURY PRESS

Mammals appeared on earth eons ago. The first mammals were land dwellers. This **bull northern fur seal** *is one of many forms of "warm blood" that have returned to the sea to find a home.*

The Danbury Press
A Division of Grolier Enterprises Inc.

Publisher: Robert B. Clarke

Production Supervision: William Frampton

Published by The World Publishing Company

Published simultaneously in Canada
by Nelson, Foster & Scott Ltd.

First printing—1973

Copyright © 1973 by Jacques-Yves Cousteau
All rights reserved

ISBN O-529-05074-9
Library of Congress catalog card number: 72-87710

Printed in the United States of America

Project Director: Steven Schepp

Managing Editor: Ruth Dugan
Assistant Managing Editor: Christine Names
Senior Editor: Rick Vahan
Research Coordinator: Elinora Yadoff
Assistant Editors: Jill Fairchild
                  Joanne Cozzi

Art Director and Designer: Gail Ash

Assistant to the Art Director: Martina Franz
Illustrations Editor: Howard Koslow

Production Manager: Bernard Kass

Science Consultant: Richard C. Murphy

Creative Consultant: Milton Charles

Typography: Nu-Type Service, Inc.

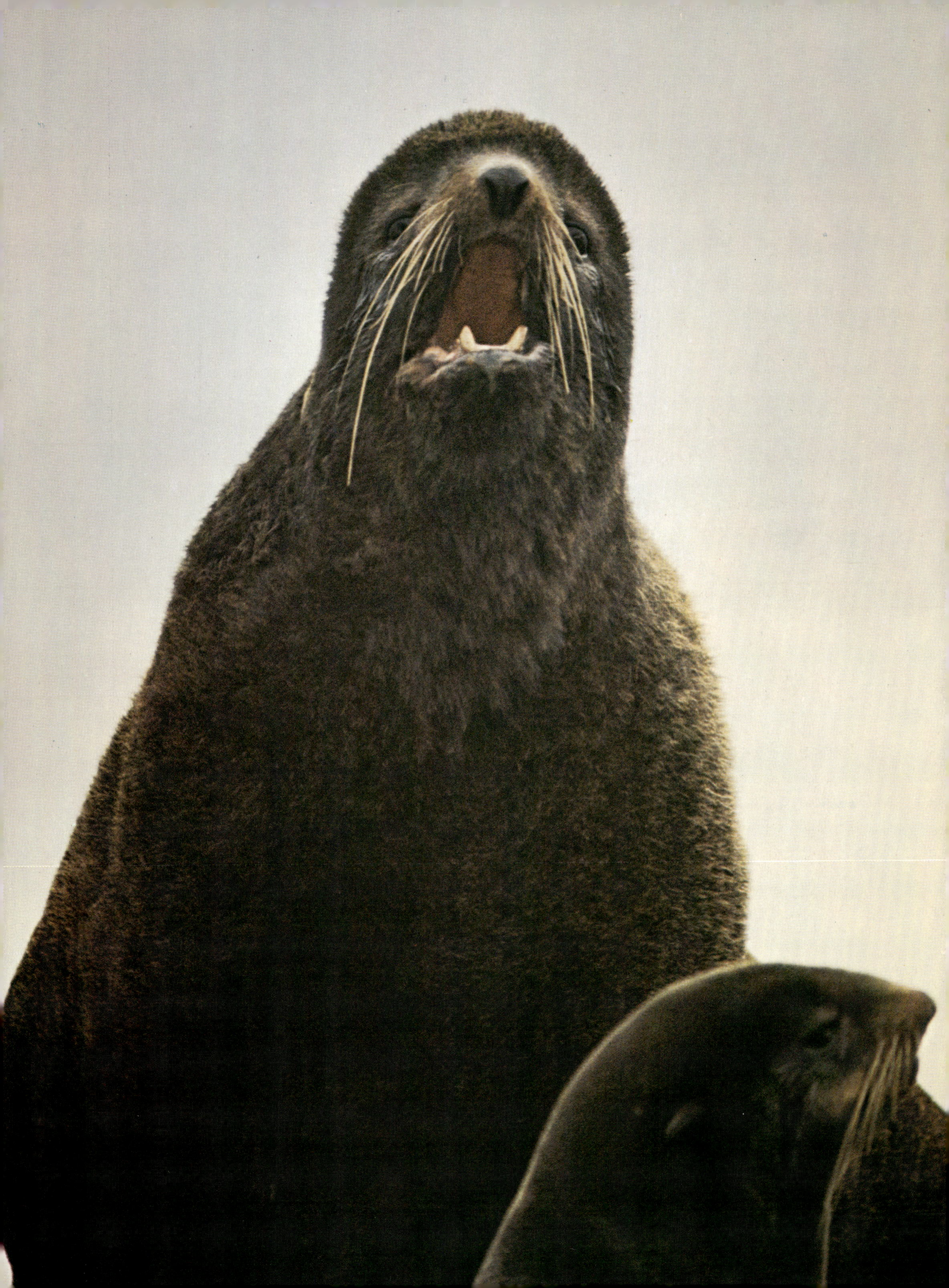

Contents

The earlier forms of life, because they were little else than "water alive," had a body temperature equal to that of the surrounding ocean. A complete line of "cold blood" invertebrates and vertebrates were mere improvements of those simple creatures. But when seabirds and marine mammals appeared, they demonstrated the supremacy of the "warm blood" system and they became the SOVEREIGNS OF THE SEA. 8

But these mammals and birds favor some particular habitat, ON LAND, SEA, AND AIR (Chapter I). In these habitats they successfully cope with the problems of survival—finding food and shelter and bearing and raising young. The sea shelters many species—some, like the cetaceans, never leave their watery home; others, like the pinnipeds, often breed on the shore; and still others, like some birds, are land-based but find food in the sea. The anatomy of every species dependent on the ocean has become adapted to a liquid environment. 10

Every animal that has taken to the aquatic environment has developed OUTWARD APPEARANCES (Chapter II) that help make life a little easier. Streamlining gives ease of movement through the water, and thick fur coats or layers of blubber provide insulation for some sea mammals. Other remarkable adaptations for life in the sea are flippers and flukes. 24

The INNER WORKINGS (Chapter III) of all these creatures have become modified and make possible their life in the sea. They have no additional or different organs than terrestrial animals. But specialized circulatory and respiratory systems enable marine animals to dive and remain underwater for lengthy periods of time without the danger of pressure-induced problems. The digestive, excretory, and reproductive systems are also marked by important adaptations. 44

The mammals and birds of the oceans have THE SENSES (Chapter IV) we have, and to one degree or another can see, hear, feel, smell, and taste. Millions of years ago sea creatures lived on land, but when they moved into a marine habitat, their senses became greatly modfiied to function in water. Perhaps the most important adaptations were in the senses of sight and hearing. The acoustic sense has become perfected to such a degree that it often becomes much more important than sight. 70

The typical life cycle of a species is called its life history. From these MARINE BIOGRAPHIES (Chapter V) we can learn much about a species. We can reconstruct the past by studying an animal's development from conception to birth; in this process each animal undergoes the stages its ancestors passed through in evolving to the present-day form. And more importantly, by doing so, we may have an idea about the general trends of evolution, and this may guide us in preventing irreparable destruction.

84

MAN, THE DESPOILER (Chapter VI), is responsible for the death of billions of living things when he overkills during hunting, destroys habitats to build, and pollutes waters through thoughtless activities. Many species of warm-blodded sea creatures have been wiped from the face of the earth or are seriously endangered by this destructiveness. Walruses, seals, and whales are only a few of the animals that have been decimated through cruel slaughter in the name of profitable commerce. In fact, in this century alone man has been responsible for the extermination of over 800 species of terrestrial animals.

100

MAN, THE RESTORER (Chapter VII), on the other hand, is seeking ways to save what is left and additionally seeks to foster the dwindling populations of some threatened species in the face of opposition from the indifferent, the ignorant, and the wanton killers. Each time a species of marine mammal has been really protected, it has made a spectacular comeback. The sea otter, the right whale, the gray whale, and the elephant seal had narrow escapes but now seem to be safe from extinction. Although protection appears to have saved the trumpeter swan, man's recognition of the whooping crane's plight may have come too late. But it is not too late to save many endangered species through greatly increased protection.

112

USE AND ABUSE (Chapter VIII) have now to be urgently checked if the sea is to be managed decently. In some cases, studies have been aimed at saving and promoting life and have been designed to increase man's knowledge of the world for the benefit of all life-forms. In other cases, the objective has been purely selfish or frivolous, based on greed or ignorance. But the public has begun to condemn the exhibition of rare animals, which may contribute to their extinction, and the slave trade that tears animals from their kin and forces them to perform tricks. Observation in the wild remains man's most valuable weapon in his fight to save endangered species.

128

In the higher forms of warm-blooded creatures, a FERVOR IN LIVING (Chapter IX) has developed and produced maternal care, play, and love.

142

Introduction: Sovereigns of the Sea

Anyone who has seen the early morning feeding at sea on a calm day has also witnessed the superiority of the warm-blooded physiology over the cold-blooded way of life. Sardines, sprats, anchovies, and other schooling fry, chased relentlessly from below by faster, larger fish such as jacks, turn the surface into a boiling kettle, while seabirds, like dive bombers, take their share of the feast. But the aggressive predator-fish—among them bonitos and bass—are themselves easy prey for the second circle of divers: the dolphins. Further away a few sharks shyly roam, waiting for the end of the feeding frenzy to scoop up the remains. They know that if they venture too close, they will be severely punished by the dolphins.

From the air or from below, the fish are no match for the warm-blooded birds and mammals. A booby can pick up a flyingfish chased by a tuna, and sea lions can easily steal the game of hunting jacks.

The warmer the blood, the higher the efficiency of the living thermodynamic machine. Birds and sea mammals have acquired their power while evolving out of the oceans, and in this book we will describe how their physiology has coped with such problems as keeping their central temperature constant or holding their breath during deep, prolonged dives. The heat-exchanging system displayed in their flukes, fins, wings, or webbed feet is a prodigy of ingenuity. Surprisingly, a few powerful, fast-cruising fish, such as the tuna, have developed a very similar heat exchange, in order to keep their hard-working muscles a few degrees warmer than the surrounding waters. But if the engineering principle used is the same, the purpose is quite different: the tuna is just a little warmer than the sea, though the temperature of the sea varies; birds and sea mammals, on the contrary, must keep their central body temperature constant, whatever the outside polar or tropical conditions may be.

Being gifted with a superior machinery, the warm-blooded lords of the sea have practically no feeding problem, although the fact that they need plenty of oxygen to heat them up obliges them to eat a lot more than comparably sized fish. Before they were chased or killed by men, colonies of many million seabirds used to pile up on islands, feeding twice a day in less than half an hour and gorging themselves to the point of being hardly able to fly. In the sea whales fill up with crustaceans in a few daily dives. Pilot whales sound to the deep layers where they swallow hundreds of squid or cuttlefish; porpoises, dolphins, and sea lions spend less than an hour a day to quench their appetite. All these spoiled children of the ocean are very particular about their food, which is one of the many reasons why sea mammals are difficult to nourish during the first weeks of captivity.

Having no difficulty in finding food, these warm-blooded animals have lots of leisure time, which explains why dolphins play almost all day, traveling for no essential reason, performing somersaults many times in the air, escorting boats or ships, outracing a cruiser for a few moments, as if to demonstrate their ability. Even creatures as clumsy on land as sea elephants are amazingly graceful and rapid in their liquid element; and when they come ashore to rear their young, they are so fat that they can afford to fast for several months.

Leisure time has been used by those sea mammals that have a large brain to develop wit, intelligence, communication, and even unnecessary feelings such as faithfulness, tenderness, and friendship. This is particularly true for those that have teeth and are carnivores. But it is not at all certain that the higher degree of intelligence originated from the flesh-eating habit—it may have its cause in the origin of the species. For example, some scientists think that the whalebone whales are descended from some extinct insect eaters such as anteaters, while sea lions came from felines, and toothed whales from an unknown branch of carnivorous land ancestors.

The physiological superiority of warm-blooded animals in the sea has severe limitations. Like complicated clockworks, they are in many ways fragile. Their high intellect produces psychological problems often ending in disaster. If the high combustion furnace of birds and mammals is easily provided with fuel (food), it must also be supplied with large quantities of combustive (oxygen from the air). This means that our warm friends must take in a lot of oxygen, and this in turn limits the duration of their dives. Some birds, such as penguins, cormorants, and puffins, and some cetaceans such as sperm whales, have modified (without radical change) their organs to perform dives of extended duration, and all toothed whales, seals, and sea lions have adapted to avoid the decompression accidents that a man suffers from while doing shallower dives.

On May 22, 1967, having met a large school of sperm whales, we managed to securely implant a strong but short harpoon in the blubber of the back of the largest male; such a harpoon does not penetrate the flesh and does not harm the animal. Attached to the harpoon was 1200 meters (4/5 of a mile) of nylon rope with a streamlined balloon at its end. The sperm whale then sounded vertically, and we could measure that he had reached a depth of at least 3300 feet. We followed the animal all day and recorded his calls to his colleagues. As long as there was daylight, the other sperm whales remained about three miles away but obviously communicated with our friend. But at dusk, when visibility was less than half a mile, they all closed in; there was a turmoil at the surface, and the big male was soon free. Somehow one of his relatives had used his jaws to tear out our harpoon and liberate him. Since then, we have seen the same maneuver in another case—the warm-blooded toothed whales are indeed the Sovereigns of the Sea.

Jacques-Yves Cousteau

Chapter I. On Land, Sea, and Air

"There's lots of good fish in the sea," wrote Gilbert in *The Mikado*. Indeed there are. In fact, in numbers and varieties of living things, the sea more than equals the land.

The cetaceans are warm-blooded water dwellers that are shaped like fish but are not fish. These mammals—the whales, dolphins, and porpoises—are born live, suckle their young, and breathe air. From the little four-foot harbor porpoise to the enormous 110-foot blue whale, these aquatic mammals have no need to come to land. These nonfish, however, must surface for air.

The seas are also host, though not total home, to the pinnipeds, which look less like fish and somewhat more like their landed mammal ancestors. They leave the sea only to breed, or they go to sea only for food (depending on how you look at the 32 species of seals, sea lions, and walruses of this order).

Measuring only 58 inches and weighing 100 pounds, the sea otter is the smallest of the marine mammals and is the only member of its family, which includes the mink, weasel, and river otter, to inhabit salt water.

The blue gray dugong and the purplish gray manatee, the two living members of the Sirenian order, complete the catalog of marine mammals. These large, sluggish creatures with stout bodies, forelimbs modified as flip-

> **"In numbers and varieties of actual living things, the sea more than equals the land."**

pers, and no hindlimbs may have been the source of the mermaid legend. One look at them, however, reveals that the sailors who saw human beauty in them had probably been at sea for too long.

Black, gray, brown, and dark blue are the basic colors of the animals of the sea. Some cetaceans have bellies or sides of lighter hues. Patches, spots, or streaks of white enliven the appearances of many species of porpoise, whale, and seal. The beluga is an ivory-colored adult after a blue gray youth. The common names of many of these animals reflect their color patterns: white-sided dolphin, ringed seal, banded seal, blue white dolphin, gray seal.

No bird, except maybe the penguin, is as at home in water as most pelagic mammals, but many birds spend their lives at sea except for one short period of the year when they come ashore to mate, nest, and raise their young. The prions, or whalebirds, the jaegers, or sea hawks, the albatrosses, and the shearwaters are among these pelagic birds.

Other bird families live in open waters just offshore, coming ashore little more than the pelagic birds but remaining closer to land. Sea ducks like scoters and old squaws and wintering loons and grebes follow this life.

Bay ducks like the goldeneye and scaup ducks remain closer to shore and seek haven in stormy weather on land or in shoreside ponds. The saw-billed mergansers, gulls, terns, skimmers, and a host of other species also occupy these inshore waters.

Each of these groups of mammals and birds favors some part of the sea or its coastline where each can find the food it is best adapted to capture and where each can best cope with the environment of deep water, ocean swells, shallow bays, or salt marshes.

The most nearly aquatic of all birds is the penguin. Although their wings do not permit flight, these **emperor penguins** *and the other 14 species of their family use their wings to "fly" underwater.*

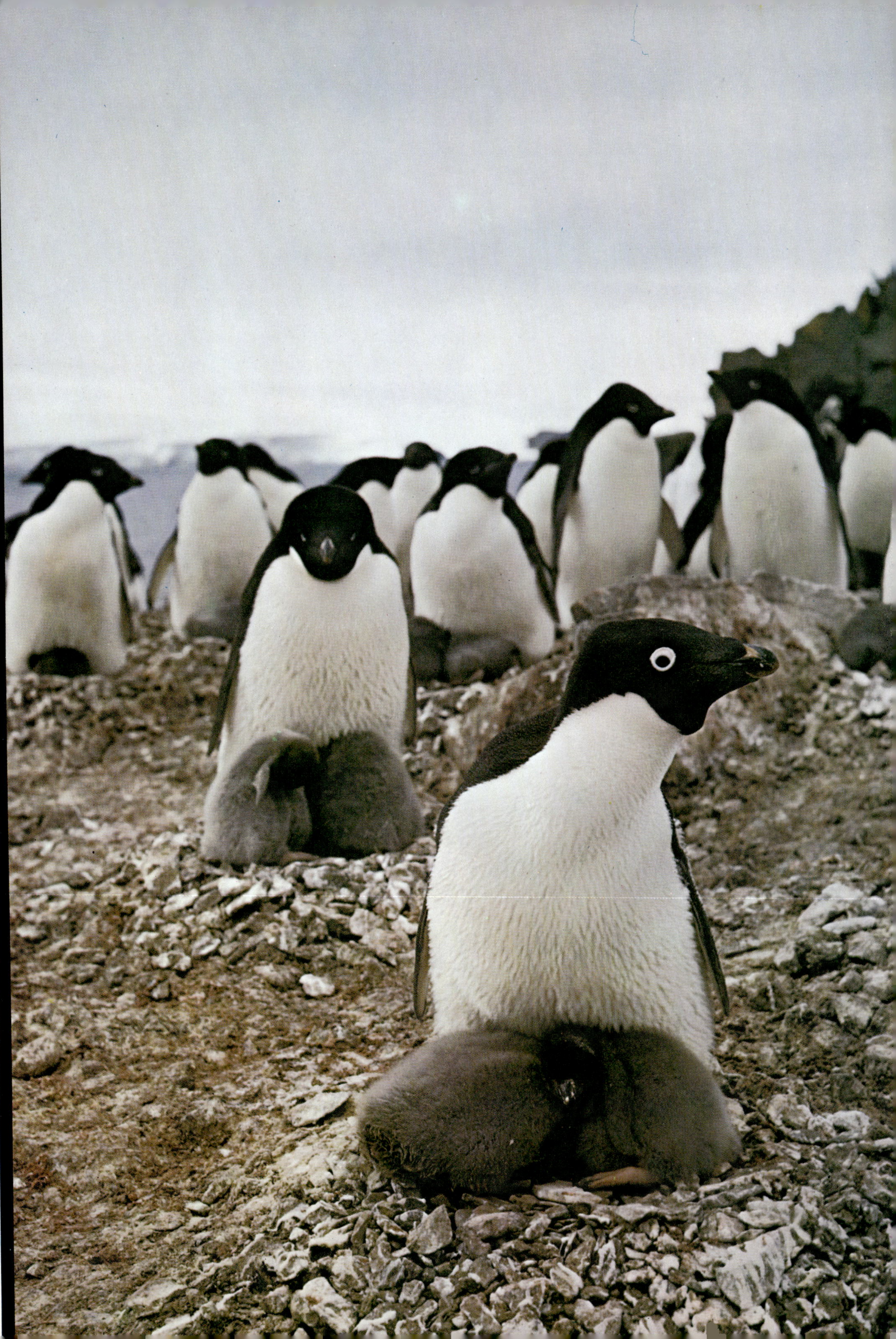

Fur seals. *Availability of food and shelter are governing factors in distribution of mammals and birds, whether oceanic or not. These fur seals prefer rocky isles in the high latitudes of temperate zones.*

Where They All Go

Two centuries and more ago, when whaling was in its heyday, a successful whaling captain had to be a tough adventurer, but also a superb seaman to navigate seldom traveled and poorly charted seas. He had to be able to predict weather in relatively unknown parts of the world, and had to know where to find the most valuable species. Knowing what whales ate, and therefore where the creatures' food lived, often led a successful captain to his prey. His harpooners and officers knew what the most vulnerable part of the whale was. These sailors of fortune, motivated by quick-gotten gains were not scientists. They were men of the sea. Today the legends and myths of their heroic struggles are viewed severely, but scientists acknowledge the perspicacity and skill of these early sailors and whalers.

With the exception of the antarctic, which is a fairly recent hunting ground, what we know about the distribution of the remaining whales today is much the same as what the whaling captains of old knew. The bowheads and the right whales were found in the arctic and antarctic regions after they had bred in the temperate zones. There are few right whales left, largely because these whales swam very slowly, were easy prey, and their natural buoyancy kept them afloat after they had been killed.

The whales are migrating animals, but they do not cross the equatorial zone, so that almost all species are divided into two different populations: the northern (boreal) and the southern (austral). The blue and finback whales spend the winter, their breeding season, in the tropics and subtropics, then move to the polar regions, where food is plentiful, for the summer. The southern hemisphere rorquals—blues and finbacks, sei and piked whales—move from the antarctic to the seas off Africa, the Bay of Bengal, the Gulf of Aden, and the South Atlantic islands of Tristan da Cunha. The northern rorquals are divided into Atlantic and Pacific groups. The gigantic blues summer in the North Pacific and go as far south as the Indian Ocean in winter, thus overlapping ranges with the southern populations but at different times of year. The North Atlantic blues move from the east and west coasts of Greenland southward along the American and European coasts in winter.

The sperm whales, which live a different sort of life because they have teeth and feed on squid and cuttlefish, live in different areas. According to Melville's classic, Captain Ahab of the *Pequod* hunted Moby Dick, the albino sperm whale, where cooler waters met equatorial currents. He chased the legendary white whale to high latitudes in summer and closer to the equator in winter.

In fact, the herds including males, females, and calves remain all year in the tropics. The old bulls, defeated by younger and stronger ones, are the only solitary sperm whales to venture to polar waters in the summer. Moby Dick, if he ever existed, could only have been a weak old male.

The little piked whale goes farther into icy seas than any other species and has been seen poking its head up through holes in the ice to breathe. Some of them get trapped too far from the edge of the pack to return to the open sea, and they die slowly of hunger.

Captains of sealing vessels had an easier job than the whalers. Seals are usually found in the higher latitudes, but there are exceptions. Sea lions reside along the west coast of North America, for example, being found from Alaska to Baja, California. At one time they were also found in Japan, but the shores where they lived were occupied by men, who destroyed their habitat; they considered seals and sea lions as fishing competitors and slaughtered them. Elephant seals live in approximately the same waters as the remaining sea lions, while harp and gray seals breed on the subarctic Canadian Atlantic coast, Scotland, northern Scandinavia, and the shores of Russia's White Sea. After the breeding season they move into the arctic regions. North fur seals breed only in the Pribilof Islands and Siberia then migrate south along the coasts where they can easily catch fish. Other fur seals live off South Africa, New Zealand, and Australia. The ringed seal and walrus live only in the arctic, while the Weddell, Ross, leopard, and crabeater seals live in antarctic seas and to the north.

Birds of the Sea

There are so many groups of seabirds that one family or another can be found in almost every part of the oceanic world. Generally, these birds can be labeled as onshore, inshore, offshore, or pelagic. Their distribution around the world is limited by geographic barriers like continents, mountains, or stretches of ocean too big for them to traverse, as well as by availability of food, presence of competing species, and weather conditions. Some birds cannot withstand extreme cold or heat, and some are unable to cope with prevailing winds.

The onshore birds include sandpipers, plovers, and their relatives. They live on sandy and rocky beaches and coastal marshes and mudflats of the temperate and subpolar regions during their breeding seasons and migrate long distances to wintering grounds in more temperate parts of the world. Their diet of small crustaceans, found only along seashores, limits them to coastlines.

Among the inshore birds are gulls, terns, pelicans, skimmers, bay ducks, geese, and swans. Their distributions is worldwide. A few, like the ivory gull, are found only in the arctic, but others, like the laughing gull, may show up on the Atlantic coast of the United States or fishing in company with pelicans along the coast of Venezuela. Pelicans, however, remain in the tropics or subtropics. Terns, near relatives of the gulls, range from pole to pole, depending on the season. Geese, ducks, and swans breed in higher temperate latitudes and winter in warmer climes. These inshore birds live within sight of land, usually coming ashore to roost each night. Their webbed feet enable them to alight in the water and swim, but they need shallower seas where they can dive and find their food.

Offshore birds, like the loons, auks, boobies, and gannets, spend most of their nonbreeding months at sea between shoreline and the edge of the continental shelf. Because the North Atlantic has extensive shelves, many of these birds are found off New England, Canada's Maritime Provinces, and the North Sea. They swim and dive in these cold winter waters then head north as far as subarctic and arctic regions to breed each summer. These strong swimmers and good divers usually hunt for fish on the continental shelves. Many, however, are not strong fliers, and they cannot get far offshore.

The pelagic birds, on the other hand, are strong fliers that spend their days aloft and roost at sea each night, except during their breeding season. Only then do they come ashore to nest and raise their young. Storm petrels, called Mother Carey's chickens by sailors, are among these web-footed swimming and diving birds with good powers of flight. These are the attributes that enable them to live at sea. The gull-like shearwaters, fulmars, and frigate birds spend much of their lives soaring over the ocean's swells. Many live on the northern high seas, but the southern half of the world, most of which is covered by oceans, has a far greater number of species and individuals. It is so unusual for these birds to come ashore that only transoceanic sailors are familiar with them. Penguins are nonflying birds that lead a life that is similar to the life of seals.

Deep water serves as a barrier for these **flamingoes,** *forcing them to remain in shallow coastal waters where they can wade on long legs.*

Shallow waters of salt marshes serve as feeding stations for shorebirds like these sanderlings.

Food, Climate, and Protection

Fishermen have known for thousands of years how to find fish. They would watch for screaming and diving seabirds flocking together and for frolicking dolphins. Where these creatures gathered, there was food.

Many animals of the sea are specialized feeders and go only where their own particular food can be found. For example, sperm whales, which feed almost exclusively on cuttlefish and squid, are especially deep divers. Whalebone whales, which feed mainly on the shrimplike krill, frequent the antarctic waters that are home for these tiny crustaceans. These whales leave the food-filled waters only to have their young in milder climates; they then return to colder waters to feed again. Rorquals eat krill and also small fish like sardines, and some herds remain all year in such seas as the Mediterranean. Dolphins, on the other hand, feed on fish, and since fish are found in almost all parts of the world's oceans and rivers, so are dolphins. The sea otter especially likes to eat sea urchins and shellfish, which are found in coastal waters from Alaska to southern California. As sea otters now begin to reoccupy the waters where they once were abundant, they head for the areas that can provide food as well as massive kelp beds. The giant kelp forests offer a haven for the sea otters from predatory orcas.

Some areas of the sea, however, are devoid of life because of natural or man-made pollution. Some pollution is made up of poisons that preclude life. Some is made up of excesses of organisms or chemicals that in themselves are not poisonous but that upset the delicate, constantly changing balance of nature. When pollutants destroy one population in an area, other populations may leave. In the bottom of the Black Sea, for example, waters are loaded with hydrogen sulphide which makes it impossible for life to exist in those depths, yet there is enough fish life in the upper zones to support a substantial commercial fishery. In estuaries and bays along the coasts of the world, where many fish live, gulls and terns abound. The salt marshes too support shorebirds, ducks, geese, and the long-legged wading birds like flamingoes, herons, ibises, and storks which feed on local fish, crustaceans, and molluscs. Pelicans ply the air close offshore, plunging like dive bombers to scoop up fish in their bills and stow them in their pouches, and frigate birds soar out over the sea, ready to pirate fish from other fish-eating birds.

Right after Christmas and well into the new year, the whale watchers turn out in numbers along the California coast. Boats of all sizes and descriptions anchor offshore within sight of land to watch the California gray whales go by. The watchers are equipped with binoculars and cameras to observe and record the southward migration of the whales to the lagoons on the Mexican coast where they give birth to their calves in a warm environment safe from predators. After not eating on their southbound trip from Alaskan waters, the whales continue their fast when they arrive in the warm waters of their nurseries. In a single season the watchers often outnumber the gray whale population of about 9000. A few years ago the sight of a gray whale was a rare one, because whale hunters had reduced their numbers and threatened them with extinction. Now, with rigid enforcement of protective laws by the U.S. and Mexican governments, the grays are growing in numbers.

Just as food attracts feeding animals, the milder climates attract the pregnant females of oceanic species. Whales and seals seek what for them are milder temperatures and safer conditions in which to bear their young and start raising them. Even those species whose young are fully capable of moving about on their own from birth, like ducks and dolphins, require some time to learn the tricks and skills needed for living, searching out and capturing food, and survival. The open polar seas are not amenable to learning the lessons of life for a tiny eider duckling or a finback whale calf. So mothers-to-be, following the instincts of their species, go to the islands, lagoons, bays, and coastal marshlands where their young have some protection until they build up the needed blubber insulation or downy undercoats to face the subzero air and near-freezing water temperatures of their usual feeding grounds with enough covering to keep them warm.

Like gray whales, northern fur seals migrate from scattered coastal areas of Alaska and Canada to areas suitable to give birth and mate. The seals, however, move northward to their nurseries in the Pribilof Islands, the Siberian coast, and the Kurile Islands. These food-rich waters offer safety to the cows while they have their pups and nurse them.

On the eastern end of Nantucket Island, off the Massachusetts coast, there is a small population of gray seals. There they live, feed, and breed, protected and in fish-filled waters. What is unusual about this group of gray seals is that they are the only ones of their species for 600 miles. Just why this relict population continues on Nantucket, no one knows, but its existence is interesting. Other gray seals occupy rocky coastal areas of eastern Canada and the Scottish coast in the eastern North Atlantic.

Animals of all kinds will travel great distances, if necessary, to find safe havens to bear their young and raise them.

*In **kelp forests** sea otters find haven from predators while they feed on abalone and urchins.*

ARCTIC CIRCLE 66°33′ N.

ASIA

N. AMERICA

EQUATOR 0°

AUSTRALIA

ANTARCTIC CIRCLE 66°33′ S.

GREENLAND

EUROPE

AFRICA

S. AMERICA

ANTARCTICA

1. NARWHAL	16. SEI WHALE
2. NORTHERN FUR SEAL	17. ATLANTIC WHITE-SIDED DOLPHIN
3. STELLER SEA LION	
4. PACIFIC BOTTLENOSE DOLPHIN	18. BLACK RIGHT WHALE
	19. ATLANTIC BOTTLENOSE DOLPHIN
5. SEA OTTER	
6. CALIFORNIA SEA LION	20. MANATEE
7. GRAY WHALE	21. DUGONG
8. MINKE WHALE	22. PLUMBEOUS DOLPHIN
9. KILLER WHALE (Orca)	23. NEW ZEALAND FUR SEAL
10. HAWAIIAN MONK SEAL	24. SPERM WHALE
	25. BLUE WHALE
11. GUADALUPE FUR SEAL	26. FINBACK WHALE
12. BELUGA	27. ROSS SEAL
13. WALRUS	28. HUMPBACK WHALE
14. HARP SEAL	29. WEDDELL SEAL
15. GREENLAND RT. WHALE	30. ELEPHANT SEAL

Where to find them. *The typical habitat of a sea-mammal is a product of all its life functions and needs. The mammal's position indicates only one point of its possible domain in the sea.*

Habitats of Sea Creatures

The type of water and shore where an animal lives is its habitat. The type of habitat it normally lives in is governed by its needs and its physical characteristics. Thus, a curlew, which is a shorebird, does not seek an oceanic habitat but looks for salt marshes and sandy beaches. There it can compete successfully with other species for food and space. Similarly, a whale, because of its physical structure, could not survive for long on a beach. Nor could a sea otter, adapted as it is for kelp forests of coastal waters, survive for long in the open ocean. And walruses, which dig clams from the soft ocean floor, would be hard pressed to make a living on a solid rock bottom. Where an animal lives, then, tells us quite a bit about it and it is important to know an animal's habitat if we are to understand much about it.

Cetaceans are usually found in a pelagic habitat, but there are exceptions. Some, like the common harbor porpoise, live principally in bays and estuaries, ascending rivers occasionally. They are seen on both coasts of the Atlantic, in Boston Harbor as well as the Westerschelde in the Netherlands. Other porpoise and dolphin species are commonly seen in the bay at Rio de Janeiro, the estuary of the La Plata between Uruguay and Argentina, Tokyo Bay and the Inland Sea of Japan, and the Bay Bengal and the river mouths on its shores. Some, like the dolphins of the Ganges River, are completely fluvial and never venture out to sea.

Other cetaceans live in specific areas of the open ocean. The lead-colored dolphin lives in the sea off India. True's porpoise stays in the warmer waters of the North Pacific, and

Gannets (right) prefer rocky sea cliffs for nesting sites where they can get to food sources quickly.

These white-sided dolphins (below) are well adapted to the open-ocean habitat.

other species occur in parts of the tropical Atlantic, Pacific, or Indian oceans. Some, like the orca, the finback, and the sperm whale, are worldwide in their distribution. There is food for them everywhere, and probably no other species can compete successfully against them in this respect. Also, they are physically adapted to live in both warm waters and cold polar oceans.

The same factors govern the distribution and habitats of seabirds. Alcids are weak fliers that have difficulty taking off from the sea's surface. Consequently, auks, murres, dovekies, and puffins spend most of their time at sea diving for fish, and only rarely do they roost ashore. To fly, they must launch themselves from the offshore rocks or paddle their feet furiously to increase their speed. Grebes too are poor fliers which flee from danger by diving rather than flying. The strongest fliers like the albatross and the frigate bird spend their days aloft far at sea where they can ride air currents.

Pinnipeds, except when migrating, stay near shore where they can find food and cavort in the surf by day and rest ashore at night.

Oceanic Migrations

Whales are the most extensive mammalian travelers on earth, barring a few humans. Because of their size, their endurance, and their reserve of energy stored in thick layers of blubber, and mainly because they weigh nothing in seawater, they can travel very economically, enabling them to migrate from feeding to breeding grounds, even though the two areas are thousands of miles apart. Finbacks spend their summers feeding in the Bering Sea and Arctic Ocean. In winter they head for their breeding grounds far to the south. Some move through the eastern Pacific to the waters off Ecuador and Peru, while others pass by the coast of Japan, past the Philippines, and into the Indian Ocean, a journey of 5000 miles which may take a month or more.

Next to cetaceans, northern fur seals are the most active migrants among mammals. They may travel 3000 miles across the Pacific from breeding territories in the north to their wintering grounds off southern California, covering the distance in less than two months. Typical of pinnipeds, they seldom are more than 100 miles from the nearest coastline. Other pinnipeds travel lesser distances, and some species are year-round residents in one locale. California sea lions range from Baja California in the south to British Columbia in the north, more than 1000 miles.

A number of bird species are migrants over impressive distances. Many traverse the distance from Canada to northern South America and back every year. The scoter, a sea duck that breeds in the arctic, migrates into northern Mexico. Canada geese that breed in northern Canada fly their well known V-formations far down into Mexico. And the golden plover wings it from Alaska and the Yukon to Patagonia in the Argentine and back every year, a round trip of 15,000 miles.

But the record for any animal, mammal or bird, is held by the arctic tern, a delicate and slender winged relative of the larger and stronger gulls. Each year arctic terns travel from the arctic to the antarctic and back—some 25,000 miles, a distance equivalent to a round-the-world trip at the equator. A one-way trip takes up to three months.

*Migrating through open-ocean waters, **dolphins** sometimes travel in large schools as these (left) or in family-sized groups at other times as they range over great distances, swimming near the surface and leaping above the waves.*

***Geese** in migration (right) travel the coastal route, very often hopping from salt marshes to bays to estuaries where they can find food. These snow geese seem to be surveying marshes below them for resting place.*

*After breeding in the low arctic, **walruses** (above) often migrate northward to slightly colder climes where food is more plentiful. Even though they are good swimmers and well insulated, walruses are gregarious and travel in groups aboard ice floes.*

Chapter II. Outward Appearances

Some 50 or 60 million years ago, for some unknown reason several species of mammals, probably quite different from those we know today, moved to the edge of the sea. They probably found food plentiful there because there were no animals big enough or smart enough to compete successfully with them. And perhaps it was comfortable for them to have their weight supported by the water. With each passing generation changes took place in their physical makeup that were

> **"Each species of warm-blooded animal that has taken to the sea has adapted in its own way."**

useful to them in their new habitat. As the changes became more functional, they spent more time in the sea until eventually they were living in the sea most or all of the time.

Piglike creatures were probably the ancestors of odontocetes, or toothed whales, like sperm whales, porpoises, and dolphins, while relatives of anteaters may have been the ancestors of whalebone whales or mysticetes.

About the same time, but not necessarily in the same places, a group of hooved animals probably related to the elephant also moved to the shallow waters of coastal marshes and the banks of river mouths. These were to become the present day sirenians—manatees and dugongs. Like the cetaceans, the manatees and dugongs have no external trace of hindlimbs, only a vestige in the form of a couple of bones imbedded deeply in muscles near the base of the tail. Their forelimbs, like those of other marine mammals, are paddlelike and ideally suited for life in the water. Their tails are large, powerful, and useful to the sirenians in propelling themselves through the water.

Long after cetaceans and sirenians had developed their ability to live in the sea, another group of terrestrial animals took to the ocean. They probably looked something like wolves or lions and took to the water some 25 to 30 million years ago. These were to become today's pinnipeds, evolving toward a similar form from two separate groups of animals. One of these groups was bearlike and developed into the eared seals and the walruses. Eared seals include the fur seals and sea lions. They developed paddlelike forelimbs and broadly webbed hindlimbs, which they use to move about on land, however clumsily. The other group developed from an otterlike ancestor into today's true seals—the gray, harbor, monk, elephant, and harp seals. They too have paddlelike forelimbs and hindlimbs, but they are unable to draw the latter forward for use on land. Instead, on land they must crawl like caterpillars and use their front flippers to pull their bodies around. These true seals appear to be more fully adapted to aquatic life.

Among the birds, animals that stemmed from tree-dwelling reptiles some 140 million years ago, the adaptation was for flight first, then for aquatic life among the waterfowl. While birds' bones became lighter in weight and permitted flight, the waterfowl kept bones slightly heavier than other birds, which helped diving in search of fish.

Each species of warm-blooded animal that has taken to the aquatic environment has adapted in its own way. Yet there are many parallels, even among animals of widely divergent ancestry, as a consequence of necessity and natural selection.

*Enormous proboscis of the **male elephant seal** gave it its name. Its purpose, not fully understood, includes resonating chamber for its vocalizations.*

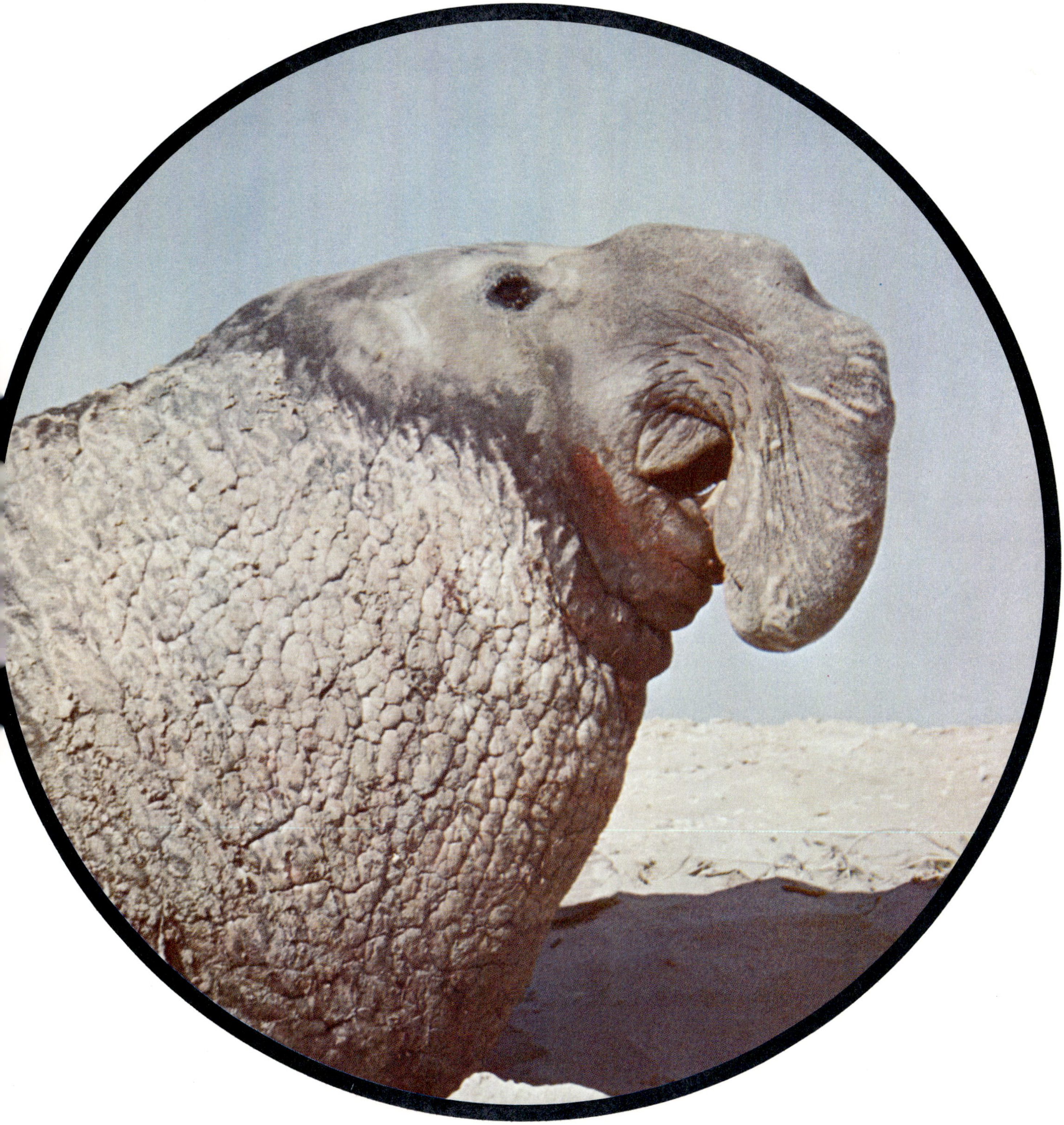

*The **gannet's** form is ideally suited for its plummeting dives and underwater swims as it hunts and pursues the ocean fish that make up its diet.*

Streamlining

Anyone who has tried to move an outstretched hand through bathwater has felt the strong water drag and can understand the necessity of streamlining. The shape of a submarine's hull is smooth, with few if any projections to offer resistance to the water and create turbulence that might slow the craft. The lines of the sub allow the water to slip past with minimal friction; less fuel is therefore required to drive it forward.

Man has copied his designs from many of the animals in the sea. Streamlining allows sea animals to move swiftly and efficiently with a minimum expenditure of energy. Dolphins, whales, and seals are highly streamlined, since populsion is the main requirement of their overall body design. Even the cumbersome-looking walrus and the strangely shaped manatee offer little resistance to the water as they swim, becoming sleek and graceful creatures in their own element. Their weight distorts their shape when they venture out of the water. The usual streamlined form is rounded at the front, widening to one-third of the length, then tapering thinner to the hind end.

In evolving toward this shape, marine mammals have undergone a distortion of their skulls in which the nostrils are pushed back until they are atop the head. This enables the animal to breathe at the surface without lunging out of the water. It need only break the sea's surface with the top of its head, open its blowhole quickly and exhale, then inhale quickly, close its blowhole, and submerge. All this occurs in two or three seconds, thanks to powerful chest muscles. In addition to the distortion of the skull, cetaceans have all but lost their necks as their cervical vertebrae have become compressed and blubber has filled in the natural constriction behind the head.

The effect of being buoyed up by water and blubber has meant that cetaceans' bones have become lighter in weight. But because of this, beached cetaceans often suffer serious injury to their internal organs. Their own weight, when they are out of water, is too much for their light bone structures.

Among birds, streamlining enables them to pass through air or water with the least effort and greatest economy of fuel. Gannets, loons, and terns are only a few of the bird families that benefit from streamlined body design. Gannets are quick strong fliers which dive into the sea from 30 or more feet above the surface to capture their food. They plummet downward cleaving first the air, then the water, as they pursue their quarry. Underwater, loons are dramatically streamlined with pointed bills, slender necks, and gradually widening bodies. The terns too, which are slender, lightweight, and sleekly trim, can fly and dive with ease. Even the ungainly pelicans are designed so they can switch from what looks like lumbering flight to dive-bomber sleekness as they pursue fish.

Sleek streamlining, like that of this **common dolphin,** *and its hairless exterior, helps give it speed as it cuts the water and cleaves the air.*

All-Purpose Coats

A bag of bones and a hank of hair describes most mammals, and if you agree that feathers are related to hair, it applies to birds as well. The bag that contains the bones is made of skin. The hair or feathers are attached to the outside for protection from temperature extremes and scrapes.

In birds, the feathers evolved from the scales of the ancient reptiles they descended from. They are a lightweight addition that gives the bird more wing area and streamlining. They are insulation that birds can fluff up to keep warm or cool. Oil glands in the skin of aquatic birds help keep them afloat, using surface tension effect. Water runs off their oil-groomed feathers. If thick, sticky crude oil escapes from an oil well at sea, it sticks to the feathers, dissolves the natural skin oil, and weighs the luckless bird down to sink and drown. A bird's outer feathers are tough enough to protect against physical scrapes. Beneath is a layer of soft down, feathers that insulate the bird.

Seals have fur, and it serves as protection against scrapes and cuts as well as insulation when they rest on shore or on the ice.

In many eared seals, the fur is thicker and has two layers. The outer layer of guard hairs is coarse and tough. The inner layer is fine and soft for insulation in air only. Sea lions do not have the full undercoat that the fur seals have for insulation, perhaps explaining why they usually live in warmer climates.

The walruses have a very short fur, but a tough skin that protects them from scrapes. The thick blubber under the skin is their sole insulation. The sort of brush they have around their jaws is very coarse and bristly, serving a sensory function like cat whiskers.

Cetaceans are a bag of bones, too, but most have hardly a hank of hair on their entire bodies. Some have a few coarse hairs around the muzzle that serve a sensory function.

Some whales have long parallel grooves on the ventral side, which run from chin to abdomen. Sperm and right whales have none. Gray whales have a few, and the rorquals have as many as 100 grooves. These grooves allow for expansion of the skin when the whale opens its wide mouth and takes in huge quantities of water before it contracts the accordion crop to filter the water out and keep whatever food there was in it. In normal swimming the grooves remain closed.

*The **fur seals'** tough coat stands out only when dry. When this northern fur seal bull (above) enters the water, its fur will become slicked down, keeping friction with water low.*

*A fur coat and a layer of blubber protect this young **elephant seal** (left) from its harsh environment.*

*Rich plumage offers protection for this adult **booby bird** (right). Its young has only a downy coat.*

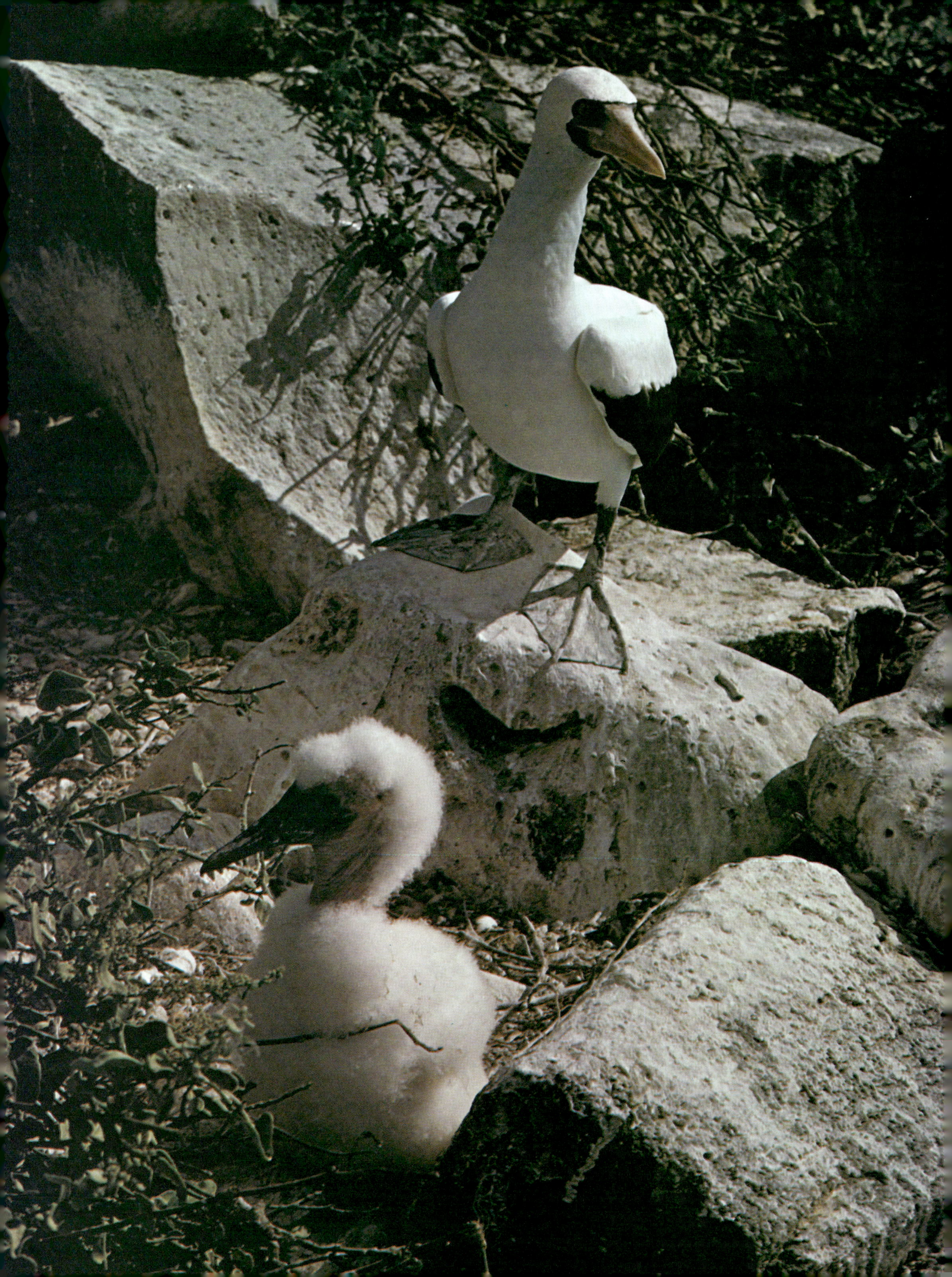

Something under the Skin

Fur, except in the case of the sea otters which manage to keep theirs loaded with tiny air bubbles, is only useful for insulation in the air. Therefore walruses and cetaceans depend on blubber to keep warm. Blubber is the thick layer of fatty tissue between skin and muscle. The amount of blubber varies with species and season. Those species living in colder waters tend to have thicker blubber than those in warmer waters. Blubber is buoyant and helps keep whales and dolphins afloat. The right whales have the thickest blubber, and this is why they float when killed. When right whales were more plentiful, whalers hunted them because they could rend so much oil from the liberal layer of fat on them. Some Greenland right whales have blubber 28 inches

Sailors on a whaling vessel peel away the blubbery coat of a freshly killed whale. In some whales and dolphins, blubber makes up more than 40 percent of their total body weight.

thick, making up more than 40 percent of the animal's total weight. In rorquals the blubber represents between 20 and 30 percent of the body weight. In the sperm whale it is more than 30 percent. Dolphins may be 30 to 45 percent blubber by weight, and porpoises have been known to carry half their weight in blubber.

The thickness of the blubber varies from one part of the body to another. The flippers and flukes have little or none of it on them and therefore help a whale control its temperature. There is only a thin layer of fat around the eyes and the rest of the head. Farther back on the body the blubber is very thick.

Blubber keeps walruses warm when swimming in ice-filled waters or lying about on ice floes. The fat falls in folds from their necks and around their chests, keeping the vital centers of the body warm.

Besides its insulating and buoyant properties, blubber is also a food reserve for the whales. In cold waters they feed upon the abundant fatty shrimplike krill and store blubber reserves. As they head from food-rich polar seas into warmer waters, food becomes scarcer so they draw on their blubber for energy. Consequently whales in the tropics have thinner blubber layers.

In pinnipeds (seals, sea lions, and walruses), the layer of fat under the skin is relatively thinner than in cetaceans and is usually about three inches thick. Male fur seals and elephant seals, for example, may go without eating for up to three months during their breeding season. When they first come ashore on their breeding grounds, these seals have voluminous folds of fat hanging from their necks and covering their chests. By the time they put to sea after breeding season, they are relatively slender. The females, during the nursing period, draw drastically on their fat reserves to produce huge quantities of milk for their calves.

Birds too have fatty tissue within them that serves as a food reserve when they are migrating. With little time to eat while heading to or from their breeding grounds, the birds use up their fatty reserves for nourishment. Duck hunters know their early season victims are full of fatty deposits, while those they shoot later in the season are not.

*A thick insulation of fat surrounds a contented looking group of **seals**, protecting them from the icy waters they inhabit. The fat is blubber, like that of whales (opposite page), and it keeps them warm when they periodically shed their fur.*

Coming to a Head

Once there was a horselike creature with a single, long horn that extended from the middle of its forehead. This mythical creature was the unicorn, and it was said that anyone who possessed the spiral horn of one of these creatures enjoyed great powers. It was believed that the horn of a unicorn would endow its owner with greater sexual prowess and would assure fertility. Just how these legends got started is not clear, but they may have stemmed from the Basque whalers who brought the tusks of narwhals back with them from their arctic whaling expeditions. The narwhal is a small whale that lives in arctic waters north of Russia and Canada. Its tusk is the left upper incisor that is one of its top teeth and grows in a counterclockwise spiral to lengths up to nearly eight feet in a 12-foot narwhal. It is found only on sexually mature males. No one knows just what the function of this tooth is. Some believe it is used to stir up shrimps and crabs from the sea floor that make up part of the narwhal's diet. Others feel it is a weapon used in ritualized combat between males. A hunter theorized it might be used to punch breathing holes in the ice. The narwhal's tusk is just one of the several unusual features on the heads of the mammals of the sea.

Rorquals, for example, have long, bony plates with hairy edges inside their mouths instead of teeth. These plates are called baleen and are made of the same horny substance as our hair and fingernails. When rorquals feed, they engulf in their inflated mouths vast quantities of shrimp along with great amounts of water. Water is forced out of their mouth either by speed, by a power-

The powerful hooked beak of the **gannet** *serves it well for tearing into fish as well as for use in defending itself against would-be attackers.*

Manatees munch water hyacinths, their favorite food, by plucking the aquatic plants with their lips and turning them into their mouths.

ful constriction of the grooved crop, or by pushing their massive tongues upward to the roofs of their mouth and thus the shrimp and other food is strained out on their baleen.

Dolphins, porpoises, and sperm whales, on the other hand, are well equipped with teeth. Some species have as many as 300, some as few as 42. They use them to hold their quarry, usually slippery fish, not to chew.

The vegetarian manatees use their lips to gather water hyacinths or hydrilla weeds that make up their diet. Seals and sea lions, with their doglike heads, are well equipped with teeth they use more for holding than for chewing, as dolphins and porpoises do. Among birds, the shapes of bills range from the huge bill and pouch of the pelican to the tiny bills of storm petrels and the long, slender bills of dowitchers and sandpipers. The broadbilled prion, or whale bird, has strainers on the edges of its bill, comparable to baleen in that the bird expels water through the strainer and retains the food.

Many marine mammals have no external ears, but this is no hindrance. Fur seals and sea lions have small external ears, however, which they lay back while swimming. Small openings mark the ears in birds and true seals. In cetaceans and sirenians, a barely visible crease shows where the ear is.

The most unusual feature on the heads of cetaceans is their nostrils, or blowholes, which have moved dramatically from the front of their heads to the top. In cetaceans both nostrils have merged into one blowhole which remains sealed during dives. The sperm whale has its blowhole located on the left side of its giant head. Other sea creatures have nostrils in more conventional locations, but they have adapted to sea life. Pinnipeds, for example, have nostrils that are closed when relaxed so they, like the cetaceans, must make a conscious effort to breathe.

*Beaklike jaws of the **bottle-nosed dolphin** are full of teeth which hold slippery fish. Dolphins and other cetaceans do not use their teeth for chewing.*

*Hindlimbs of a **sea lion** still show digits that once were toes. Although this mammal has gone back to the sea, the internal structure of its limbs are still similar to its land-dwelling relatives.*

Hands Become Flippers

Although among either quadrupeds or superior apes there is little difference between feet and hands, no one ever thinks of our hands as front feet, but that's what they are. A look at X rays of hands and feet is convincing proof. Among the mammals which live in the sea front limbs have become broad, flattened appendages. The X rays once again prove the point, for within the hands, paws, and flippers of so many mammals, the structure and arrangement of the bones bears a striking similarity.

Cetaceans have within their flippers, or pectoral fins, a series of bones that compare with the bones of our wrists, hands, and fingers. For hindlimbs, they have little or nothing, because over the millions of years they have spent adapting to the aquatic life, they have all but lost every vestige of hind legs. Some cetaceans have no trace of them, and some have only vestigal limbs and hipbones buried deep within their bodies. In adapting to life in the sea, however, whales and dolphins have added something that might pass for a new limb. Their tails have become broadly flattened with a large area that can push great quantities of water behind them. These are the flukes which, with the powerful body muscles, propel these streamlined creatures so swiftly through the sea.

Among pinnipeds (a word that means "fin footed") the front limbs are flippers. In the case of sea lions and fur seals these flippers have grown almost to look like the wings of swallows, and they use them very much indeed as swallows use their wings. The hindlimbs are flat and broad, sometimes connected with each other to give a greater surface with which to push the water. But in sea lions, in front and hind flippers, the structure of wrists and fingers is visible through the skin, and nails protrude at the end of each hidden finger. In walruses both front flippers and hindlimbs are less developed.

In sirenians the front limbs are short, flat, and broad like paddles and are used chiefly for steering and stabilization. The hindlimbs have disappeared, leaving only a broad round tail that is used like a whale's fluke to supply power for forward motion.

*The swift and graceful movements of the **sea lion** result from front and rear limbs that have been modified for increased mobility underwater. Their digital arrangement resembles our hands and feet.*

Aquatic Limbs

Sea otters are the least changed from terrestrial form of all the marine mammals, but their limbs too are beginning to show signs of flattening and growing webs between the fingers. Sea otters use their flattened hind legs for swimming and their tail has grown and has flattened to become a third hind flipper. They use their forelegs, with their well-developed fingers, for grasping food. Unlike any other marine mammals, they still have well-developed, retractable claws, which they use in food gathering.

Beavers are not marine creatures, but they are aquatic, have developed webbed hind feet, and a wide flattened tail covered with scales. The webbed feet are used to swim, but the flattened tail, though also used to increase swimming speed in emergencies, is primarily a tool used to excavate or to cement mud constructions.

Another freshwater, warm-blooded creature is Australia's platypus, a strange mammal with a duck's bill and short webbed feet used to swim and to dig.

Polar bears are the least aquatic of the marine mammals, obviously not as well adapted to life in the water as whales or seals. They walk and run easily on snow, ice, or over bare ground on sturdy legs and huge paws. However, since the polar bear's diet consists mainly of seal, the bears spend a great deal of time in water. They enter water rather reluctantly, easing in backward, one hind leg following the other, grasping the ice with the short claws of the front paws until only the nose protrudes. In water the polar bear is a strong, graceful, almost indefatigable swimmer, capable of attaining speeds of four to five knots. Polar bears sometimes use all four legs as oars, but most often they dog paddle, stroking with their powerful webbed front paws and using the trailing hind legs as rudders. While stalking a seal, a polar bear is able to duck under water or ice for about two minutes at a time. It is capable of diving to a depth of from three to six feet and can make spectacular leaps from water onto the ice.

If man were to adapt to life in the sea, he would undergo physical adaptations over thousands of generations. We might develop broader, flatter feet with webbing between our fingers and toes. Our nostrils would be able to open and close; our eyes would adapt to undersea vision; our blood would alter its volume and chemical composition; we

Sea otter's limbs (below) have just begun the gradual change from terrestrial form to the more flattened shape that enables them to move about in water. Their clawed digits function almost as fingers.

Polar bears *(above) move over snow and ice and through water with startling speed and agility. Their well-developed limbs provide them with the mobility they need to hunt and capture food.*

would cover ourselves with a thick coat of fat and improve our streamlining. Whatever changes might come about, our skeletons would likely remain recognizable when compared to our bone structures of today. This is approximately what happened among those animals that have adapted to sea life over the past 20 to 60 million years.

Flippers and Flukes

Far at sea, where the only sound is usually that of wind and waves, there occasionally come loud reports that resound for miles. These are the sounds of sperm whales or of humpback whales breaching.

Humpbacks breach mainly during the mating season especially during the courting that goes on for weeks in tropical waters. As far as sperm whales are concerned, the crew of the research ship *Calypso* has repeatedly observed large whales clearing the water, their center of gravity reaching about 18 feet in altitude, and then falling back flat on the surface, raising huge geysers of seawater. Mathematically such leaps are only possible if the vertical speed of the whale reaches 22 knots before breaching. On board *Calypso* we believed this happened when a sperm whale found itself short of breath after staying too long at great depths during a fight with a giant squid or for any other reason, and it then sped desperately upward for fresh air.

At other times whales stand on their heads in the water with only their flukes out, waving back and forth. Sometimes when a herd is disturbed by whalers, the bull in charge of the community emerges vertically until its eye is out of the water, which means about 25 percent of its weight has to be supported by a constant thrust of its beating fluke. It stays in this position 30 to 60 seconds, turning around on its axis to study the situation, then it sinks back into the sea, probably to give instructions to the group.

These powerful flukes are rightly feared by whalers. Some whalers have been thrown overboard by the tails of harpooned sperm whales, which out of anger or fear, or more probably just by trying to escape, have shattered whale boats and the men in them with mighty swipes of their powerful flukes.

Dolphins and orcas are swimming machines, as advanced in the warm-blooded category as tunas are in the cold-blooded family. Their flukes are perfectly coordinated with the powerful muscles of their streamlined body and the number of fluke beats is almost linearly proportional to speed—it is comparable to a ship's speed being a direct function of propeller revolutions per minute.

All these toothed cetaceans can leap out of water or somersault for play.

Cetaceans' front limbs are in the form of flippers like most other marine mammals. Flippers of whales are only used as rudders or planes to make short turns or to avoid obstacles. The flippers of humpback whales have almost disproportionate size, but account for the playful, unpredictable maneuvers of the animal. Flukes, on the contrary, are almost exclusively a powerful propelling system enabling whales to gather respectable speeds. Sperm whales can, when irritated, reach speeds of 12 knots and more! One reportedly towed an Azores fishing boat at a speed of 20 knots. The enormous blue whale can get up to 15 knots for long runs and 20 knots in sprints. Finbacks have been clocked at 18 knots, while the humpbacks and gray whales can do at least 15 knots. The record seems to be held by the sei whale which can make 35 knots in short sprints.

The most original use of flippers is demonstrated by sea lions and fur seals; it is with powerful strokes of their long, thin, front flippers that they "fly" underwater, while the soft relaxed hind fins are mostly used for steering. Sea lions are not quite the fastest, but probably are the most maneuverable marine mammal. Harbor, monk, and elephant seals swim only with body undulations: their hind flippers are stretched like two parallel caudal fins, beating *laterally* like those of fish. The walrus makes little propulsive use of its limbs, but its thick body is extremely flexible and is rhythmically bent to become an efficient traveling wave.

The flattened limbs and finlike tail of sea otters enable them to swim well. They can also move about on land on all fours, although their aquatic life has left them weaker than their more terrestrial relatives, the minks and weasels.

Awesome flukes *are all that is visible as a mighty leviathan submerges. Whales and dolphins use flukes as a means of propulsion. Flippers are employed only as steering mechanisms.*

Hands Become Wings

A bird's wings are its front limbs, analogous with our arms, and they probably developed from the front limbs of tree-dwelling lizards some 140 million years ago. Among all the mutations that changed the lizard in the course of centuries, some improved their ability to leap from tree to tree. Eventually their limbs became what we now call wings: they had become the first birds. Wings are the organs of flight for the albatross, sustaining it for flights over the sea. They carry the frigate bird for long hours in search of prey. But the aerodynamic design of wings is mostly dependent on feathers. The naked body of a bird seems less made to fly than that of bats, for example. Feathers, which are an exceptionally light, strong, and flexible material, fill in the wings and give them the various curvatures and shapes needed to take off, fly, or land. In many seabirds, however, flight is not so important, and consequently they are not strong fliers. Alcids, loons, grebes, and coots have trouble getting airborne. Penguins and the flightless cormorants of the Galápagos Islands are unable to fly at all and have adapted to a more thoroughly aquatic existence than any other birds.

Sea gulls (below) feed on the remains of a dead whale which is beneath the surface. A brown gull attacks those below in hope they will release their food for him to retrieve from the surface.

The hindlimbs of birds, their legs, are also often specialized. Birds that developed webbed or lobed feet had advantages in water and became aquatic. Swans, ducks, and geese have webbing between their toes.

Grebes and coots have lobes on each of their toes, and some shorebirds have half-webbed feet. Probably the most spectacular development of legs among aquatic birds, however, is the long stiltlike legs of the herons, storks, ibises, and flamingoes. These limbs are so thin they seem incapable of supporting weight.

Sanderlings (below) scurry along sandy beaches at water's edge on tiny feet. Some other sandpipers have half-webbed feet for running over soft coastal marsh mud where they also find food.

A fulmar (above) comes in for a landing, webbed feet extended and wings flapping in short, shallow strokes to control its descent. Fulmars are pelagic birds that come ashore chiefly at breeding time.

Flight in Air and Water

Most diving birds use their wings to fly underwater in much the same way they use them to fly through the air. A few others, on the contrary, fold their wings when entering the water realm, and paddle very efficiently with their stretched legs. The body is also stretched out in order to be thinner and to offer least resistance to propulsion. Some diving ducks are in this category. Alcids, mergansers, and puffins pursue fish in the way many birds chase insects through the air. But

Cormorants swim underwater almost as well as they fly. Their streamlined shape helps in both environments. Using webbed feet to steer, they propel themselves forward by short flaps of wings.

because water is 800 times denser than air, a bird's wings must be stronger to fly underwater. Therefore, diving birds have developed heavier bones than those that do not go underwater. In most cases, the underwater fliers are not strong fliers above the surface, possibly because of their heavier construction.

The hindlimbs or legs of birds that swim, such as ducks and geese, have similarly become heavier and stronger than those of birds that spend much of their time in the air or perching.

Storm petrels, called Mother Carey's chickens by sailors, have combined an aerial life with an aquatic one. They spend much of their time flying just above the surface of the sea, dipping their webbed feet into the water and dabbling them along the surface, perhaps in search of planktonic food. Because they appear to be walking on the water when they behave this way, they have won the name "Jesus bird" in many places.

*This **rhinoceros auklet**, a member of the alcid family, and numerous other diving birds often use their wings to "fly" underwater in pursuit of fish, which represent the mainstay of their diets.*

Rails, coots, and a few species of shorebirds with lobed or semipalmate feet can walk over marshes and soft mudflats or quicksands along the shores where they live. Some, like the sandpipers and plovers, run quickly before incoming waves and fly up to avoid getting wet. Others, like the rails, use their big feet to step carefully on the soft ground, flying only rarely.

43

Chapter III. Inner Workings

If a man equipped with an underwater breathing apparatus dives deep enough for a long enough period of time, he may suffer the effects of caisson sickness, or "the bends," a painful and sometimes crippling or fatal effect of exposure to a high pressure environment. When a whale dives just as deep or deeper and for as long a period of time, it has no such problem. The reason is that the whale, over millions of years, has undergone certain physical changes. Other cetaceans and pinnipeds have undergone similar changes in their physiology, making it possible for them also to make long, deep dives.

The effect of this ability to dive deep and long has enabled marine mammals to extend the area of the ocean in which they can obtain food. In Chapter I we saw that whales

> **"The organs of sea creatures, found in birds and mammals, have been modified to suit their needs."**

are known to dive well beyond 1000 feet beneath the sea's surface and some of them may go as deep as a mile. In these great depths as in the shallower waters, they can find squid, fish, and concentrations of various organisms upon which some species feed.

While deep diving serves these sea creatures well, more adaptations are needed for life in the sea. An aquatic existence means living in a sphere of entirely different physical and chemical makeup than that of the air where most birds and mammals make their homes. Therefore further adaptations have been made to enable these animals to cope with this vastly different environment. These adaptations include the ability to withstand the powerful heat-draining effect of the water, although the temperature variations they encounter at various levels in the sea are less important than in air. Other adaptations are specialized kidneys, respiratory and circulatory systems, and bone structures, as well as the external changes already noted.

Nowhere in any of these aquatic creatures are there any additional or different organs to aid their life in the sea. Instead the properties of the existing organs, found in all birds and mammals, have undergone modifications to suit the needs. The results of these modifications are predictable. That is, when they occur in terrestrial animals, we can anticipate what these changes will do for them. When a seal dives, its heart rate slows dramatically. We find that similarly when a man dives and holds his breath, his heart also slows. It does not slow down as dramatically as the seal's, and the exercise of diving may offset the slower heartbeat somewhat, but it is reduced significantly. The process in seal and man is directly related because both have the same basic physiological systems.

Birds too must adapt to life in the sea. The high salt content of seawater requires them to cope with the effects of the chemicals in the ocean. Like some marine mammals, some seabirds need a third eyelid, or nictitating membrane, to protect their eyes and to help them see clearly while they are diving in search of food. They must remain immune to the effects of constant motion as they ride the waves for hours and days at a time. And they require an astonishingly accurate navigational sense to migrate over an environment where there are no landmarks to guide them on their long overwater flights.

Right whales *have big, highly arched lower jaws, unique among cetaceans. Within those jaws are the baleen plates with hairlike appendages that strain plankton, the whales' principal fare, from the water.*

45

Need Amidst Plenty

>Water, water, everywhere,
>Nor any drop to drink.

Samuel Coleridge wrote these lines in *The Rime of the Ancient Mariner*. This is precisely the problem facing the warm-blooded creatures in the sea. Water is one of the key substances required by all living things, but the salt in seawater makes the oceans undrinkable in any great quantity for all mammals and birds. The salinity of the body fluids of birds and mammals is substantially lower than that of sea water. Salts draw water to them, and this includes the water in the tissues of animals in the sea. If there was no way of controlling this potential outflow of water, neither mammals, birds, fish, nor reptiles could exist in the oceanic environment. All their body fluids would soon be equalized in salinity to the sea round them.

First in importance for living in a salty environment is a source of fresh water. Fish-eating dolphins get much of their water from their food because fish flesh, though not pure fresh water, is less salty than the sea. Orcas, feeding on other mammals and birds, obtain water from the tissues of their victims. But both groups take in small amounts of seawater as well as the fluids from the flesh of their prey, which are also not fresh water. Their kidneys have to excrete the excess salt.

Seals and sea lions reportedly take in no seawater with their food, which is mostly fish. Many whales, the walrus, and the sea otter, however, eat invertebrates whose body fluids are close in salinity to seawater. One of the principal reserves of water for marine mammals is their fat or thick layers of blubber. When this fat is "burned up," or metabolized, water is a by-product.

Walruses (left) and *pelicans* (below) can survive on water released from their food. *Manatees* (right) can travel to sources of fresh water.

Marine mammals conserve the water they gather by various means. Cetaceans have no sweat glands, and because they inhale and exhale in humid atmosphere so close to the water's surface, they lose little water through their lungs while breathing.

Larger, more efficient kidneys help marine mammals dispose of the excess salts they take in. Since the kidneys of some birds are not as efficient as mammalian kidneys, a pair of glands near the nasal passages in the head help secrete excess salts. These salt glands are found in all birds, but are not always functional. (Marine reptiles, especially the Gálapagos iguanas that feed in the sea, also have salt glands.) The duct from each of the salt glands carries salt solution into the nasal cavity and thence to the outside where the solution drips off the end of the bird's beak. The system works only when excess salts accumulate, usually after feeding at sea. It has been shown in experiments that a bird can eliminate two-thirds of the salt it has ingested within four hours of feeding. In these experiments its kidneys were measured as one-tenth as efficient as the salt glands in eliminating excess salts.

47

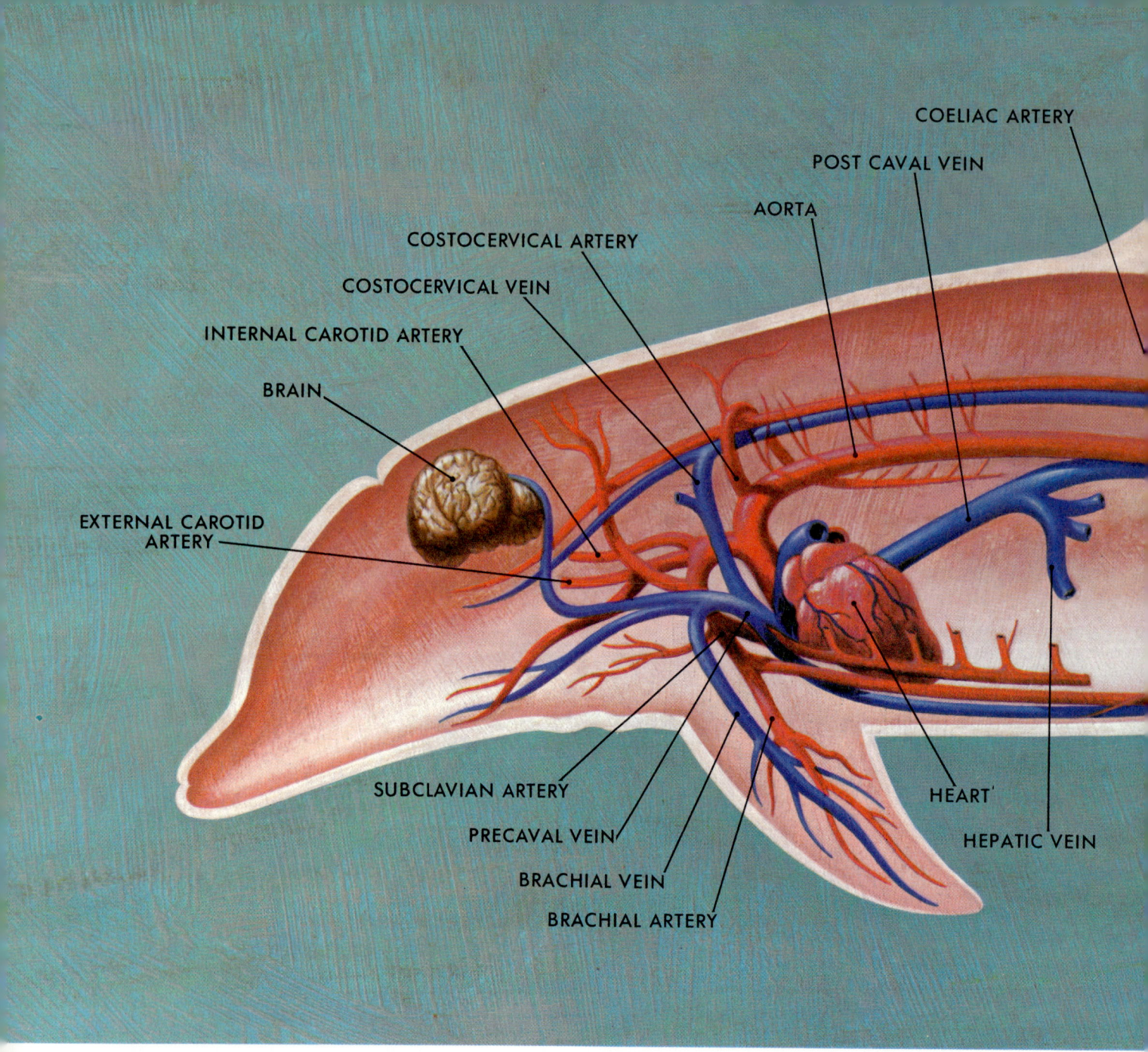

The Heart of the Matter

Before elephant seals were protected by law, they were butchered on the beach during their mating season by whalers returning home after a poor trip who tried to compensate for their losses by getting whatever fat they could quickly get.

The slaughter was disgusting and we will not describe it here. But the first seal hunters were amazed at the quantity of blood in elephant seals—105 gallons! By comparison, humans have three gallons, which is proportionately about the same quantity for an equivalent body weight. But the elephant seal has almost twice as much as man per pound of muscle, because blubber, which accounts for 40 percent of the seal's weight, is very poorly irrigated. To pump all that blood throughout its system, the elephant seal has a very ordinary heart, considering the animal's size. Cetaceans also have hearts that are only proportionate to their size. Surprisingly there is nothing extraordinary about their hearts, although their other organs are highly specialized. The usual horizontal posture combined with the apparent weightlessness of marine mammals in the sea impose a far lesser effort on the pump—the heart. The most remarkable part of the circulatory system of cetaceans is the network of tiny blood vessels throughout the fatty tissues. This network, called the retia, is made up of arteries and veins. There are a number of speculations on the purpose of the retia in whales, dolphins, and porpoises. Some physiologists have speculated that the retia's most likely function is to alleviate pressure changes during diving. But this is meaningless, since all body fluids are automatically in pressure equilibrium, and the only parts of the body that suffer from pressure changes are the air filled cavities. Some of the retia blanket small areas around the brain and the pelvic region, including the reproductive organs. The main function of the retia seems to be to store useful quantities of blood that is needed during dives.

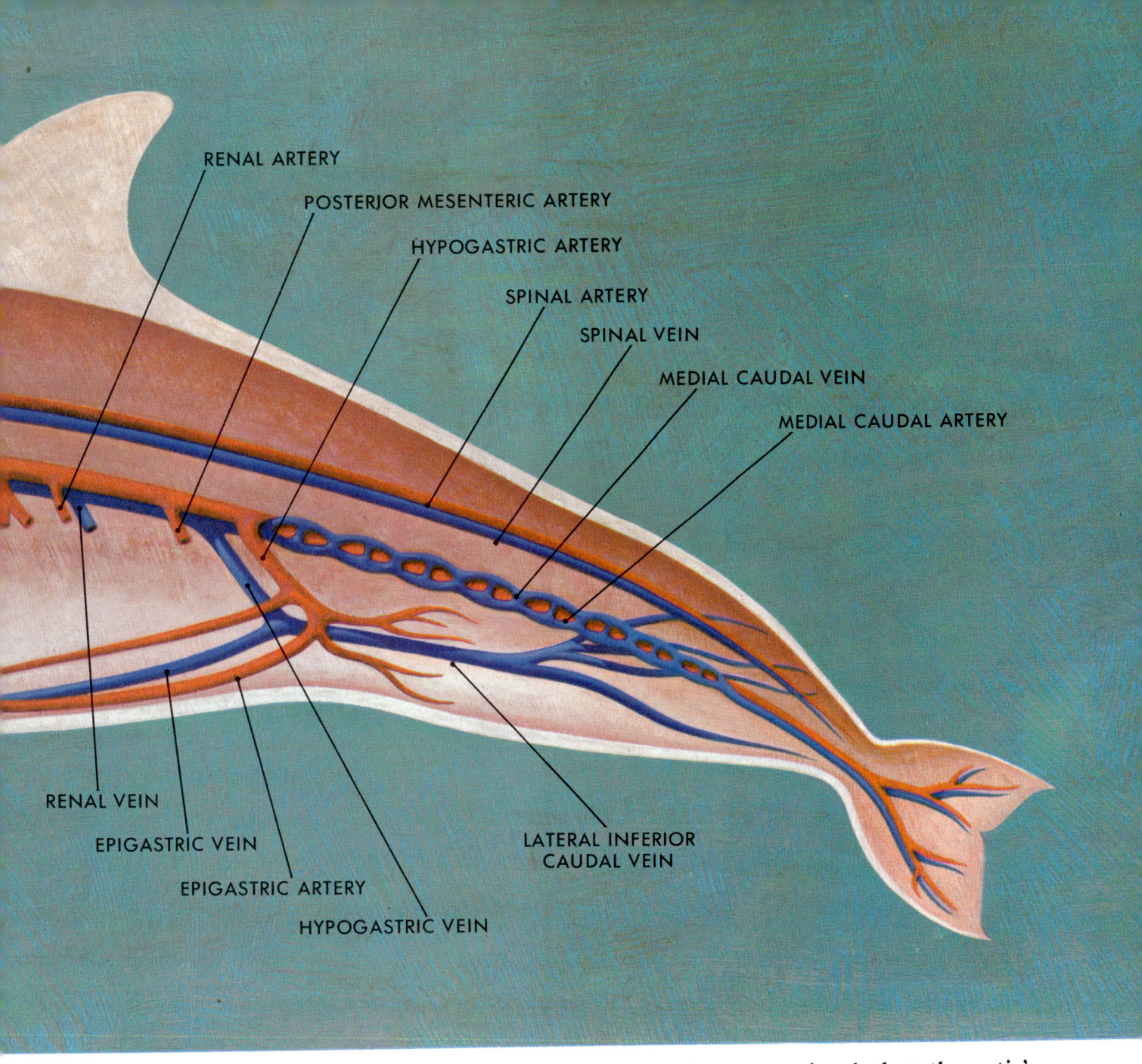

*The **streamlined shape** of cetaceans like these dolphins is one of several factors that allow them to control their body temperature and keep themselves warm although swimming in frigid seas.*

Keeping Warm/Cool

During World War II researchers sought with very limited success to devise ways of saving the lives of torpedoed sailors and airmen shot down in the North Atlantic where water temperatures ranged from 28° to 35° F. in winter. Men who had escaped injury when their ships were sunk were rapidly dying of exposure when subjected to these extremely low water temperatures. One avenue of research was to learn how whales, living in near-freezing seas, could withstand the cold. It was already well known that part of the answer was that whales have a layer of insulating blubber between the skin and muscle tissue. Researchers wanted to know, however, just how effective that insulation was. To learn this, they had to know the normal body temperature of a whale. It was extremely difficult to obtain this information.

To take a whale's temperature, you must first have a whale available. To get a whale, the researchers had to chase one, but speed of flight raised animal's temperature to an abnormal level. One group, after a long and intense chase, harpooned and paralyzed a sperm whale and then took its temperature. It was 100.7° F., far above what has since been shown to be normal. Other researchers tried shooting dart thermometers into whales in the open ocean and in oceanariums. The results showed some dolphins with central temperatures of about 100° F., but generally cetaceans have average temperatures around 95° F. Normal human body temperature is 98.6° F. and that is considered low among mammals. Cows, for example, have a normal temperature of 101.3°, that of a healthy cat is 102.2°, and that of goats normally runs 103.1°. Birds too are consistently over 100°. Among pinnipeds the fur seal was highest at 100°, but the walrus ran around 97° and the elephant seal was lowest among all pinnipeds at 93°. Only a few very primitive terrestrial mammals have normally lower temperatures.

We know that water conducts heat away from the body much faster than air does. In all cases, the central temperature of cetaceans was much higher than that of the water, which rarely reaches 85° at the surface in the tropics and drops below 32° in polar seas. Warm-blooded animals, like birds, seals, and whales, have a better thermodynamic machine than cold-blooded animals and take advantage of it in speed, energy, and endurance. But the handicap is that within narrow limits that high central body temperature has to be maintained practically constant. A few degrees more or less may mean disaster.

Research in the physiology of thermoregulation is still in its infancy. It has led to a number of imaginative but unfounded theories, and recently to a real breakthrough. The problem is, specifically, to understand how marine mammals keep themselves warm enough even in polar seas and at rest or, just as important, cool enough in the tropics when swimming hard. In the rorquals, some theorize, the ventral folds help regulate their temperatures. The bottom of each two-inch-deep fold is rich in small blood vessels. With the folds closed, there would be little loss of heat. If the folds are open to the sea, however, cold water would carry that heat away, cooling the whale. Unfortunately for this hypothesis, sperm or right whales do not have such folds but regulate their temperature just as efficiently.

Harbor seal. Besides the layer of insulating blubber that seals and other mammals have, the harbor seal has a rich supply of blood to the skin that can be used to cool it or keep it warm.

Thermoregulation

It it now well established that in whales, porpoises, and dolphins, as well as in seals, sea lions, walruses, and such birds as the penguin, the main role in the complex system of body temperature regulation is played by flippers or front limbs and by flukes or hind limbs. These flattened feet, hands, or fins are relatively thin sheets of flesh with no blubber, but they are abundantly irrigated and they allow an important cooling of the blood circulating within them.

Most of the time this cooling could be disastrous. An important reduction in the local blood circulation would reduce the heat loss, but such a primitive system of valves as the arteries would not solve the problem. Instead, a remarkable arrangement of blood vessels allows a reduction of heat loss to practically zero, while it assures irrigation of a limb that is very close to water temperature with blood that arrives there warm and returns to the general system warm.

It took man centuries to find a solution to a similar problem, faced each time one needs to circulate a fluid in a high-temperature enclosure and avoid the waste of energy that occurs when that fluid is cooled as it flows out. Such a case is found in desalination plants. To solve this problem, engineers invented the "countercurrent heat exhanger." Or rather, they thought they had invented it. In fact, it has been used in the flukes and flippers of cetaceans for millions of years.

Each of the arteries that feeds the flippers and tail with warm blood is surrounded by veins that join to form a sheath through which the blood returns to the heart. The blood in these veins, cooled by having circulated in the cold limb, is warmed by a transfer of heat from the blood in the artery which is cooled in the process. The venous blood is progressively warmed on its way back to the body and the overall heat loss is very small—the flipper or the fluke is irrigated and oxygenated by arterial blood that is almost cold. If the radiator constituted by the fin is needed at full efficiency, the cold venous blood bypasses the "heat exchanger" and returns through a network of other veins close to the skin that have no insulation.

The rest of the body is insulated from the cold surrounding water by the thick coat of blubber described earlier. This heat insulation can be further enhanced by control of the peripheral blood circulation which is automatically reduced during each dive.

This insulation is more efficient in the case of large whales than of small ones, because it is dependent upon the surface of the animal while its heat production is proportional to its weight; and the surface increases with the square while weight increases with the cube of the length. The protection is better in the case of bulky animals, like the right whale, rather than in elongated animals like the sea lion or the sei whale. In the case of sea lions or humpback whales, the large surface of the flippers is a handicap for protection but an advantage for cooling, even with the presence of the efficient "countercurrent heat exchanger" and the "thermostat" that has been described.

Seals also use their body surface as a heat exchanger. A seal's skin is thicker than a human's and is liberally supplied with tiny blood vessels that can be opened or closed. When they are open, the blood in them warms the skin and the heat is carried off by cooler water. When the vessels are closed, the seal's body heat is retained within. In cold water, a seal's skin temperature may be as low as 35° F., while its internal temperature is around 99° F.

The smallest and the least adapted of the marine mammals, the sea otter, is unable to protect itself against cold with its fat alone, but has developed a way to use the marvelous insulating properties of a fur coat that is constantly reloaded with thousands of tiny air bubbles. This the otter maintains by rubbing its fur with its paws whenever it rests on the surface; when the otter dives, only the surface of its fur gets wet, and it travels enveloped in a coat of air during dives that are not very deep or longer than several minutes.

Birds generally benefit from the same air-coat protection brought to the ultimate degree of perfection as their greased feathers are even better as a shield against cold than hair, and they are able to control individual feathers in positions that allow precise degrees of "ruffling" to imprison air either before or after a dive.

The bird's feet, especially if they are webbed, can play the role of radiators to cool off the animal if needed. Protecting them against cold is done by simply retracting the feet within the belly's feathers.

Puffins spend a large part of their lives underwater. Penguins, unable to fly, spend at least half of their lives in water that is often below the freezing point, and they travel hundreds of miles at sea, diving deep, often, and for long times. Their wings have become shorter and stronger and are built very much like sea lion's flippers. They use them as heat exchangers exactly as whales do and have independently developed a thermoregulatory system that is remarkably similar to that of cetaceans.

Blood circulation. Most marine mammals, including this crabeater seal of the antarctic, can close off blood circulation to certain parts of their bodies in an effort to control body temperature. Seals in particular use their body surface, as well as their flippers, as a means of heat exchange. Seals can regulate the flow of blood through their thick skin and in this manner control the amount of heat lost to the cold environment. Remarkably, the arrangement of blood vessels can allow a reduction of heat loss to practically zero.

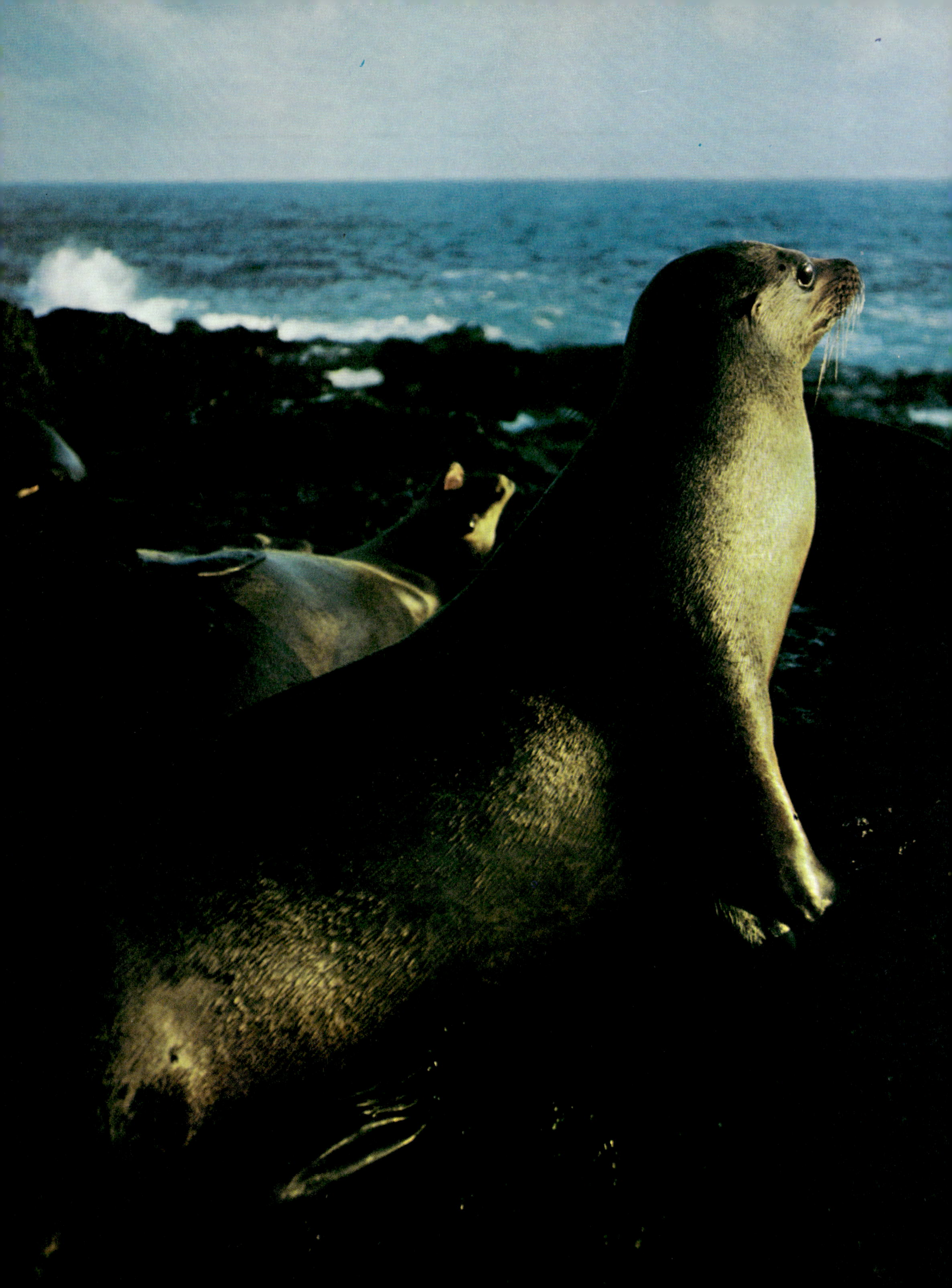

Rocks in the Stomach

Birds and mammals of the sea swallow their food whole. None is equipped for chewing. Instead, whole fish, squid, shrimp, and other prey are dissolved by gastric juices and possibly ground up by gravel and stones in the stomach or crop. Even the orca, with its powerful teeth, only tears its large prey apart, swallowing large chunks of the victim.

Cetaceans have a three-part stomach reminiscent of cattle and other cud-chewing animals. The first part is actually a great widening of the esophagus, which is the food tube between mouth and stomach. In this first part many whales have stones, or gastroliths. The muscular contractions of this first stomach with its rocks helps grind the large chunks of food therein. The second section of the stomach is comparable to the human organ and secretes hydrochloric acid, pepsin, and other digestive enzymes. The third stomach is mostly smooth-walled but has a few glands that secrete digestive juices. Rorquals can hold up to a ton of krill in the first two stomach sections. From these stomachs, the food passes through the small and large intestines, which in cetaceans are quite long.

Pinnipeds, like cetaceans, often have stones in their stomachs. Again, while the function of these rocks is not clearly understood, most scientists believe their function is similar to the gravel in a bird's crop—for grinding large pieces of food into smaller ones. One of the most puzzling features of the seals' digestive systems is the length of their small intestines. Carnivorous animals—those that eat flesh—usually have short intestinal tracts, while vegetarians like cattle have long tracts. A two-foot-long dog, for example, has an intestine measuring no more than 10 or 12 feet. A six-foot man's gut measures 23 feet, and a six-foot-long cow has more than 100 feet of intestines. This being the case, scientists would like to know why a 15-foot elephant seal, a fish-eater, has a small intestine measuring 660 feet. The question of the rocks in the stomach also remains unanswered, because while they may be for grinding food, they could, as in the crocodile, have to do with buoyancy. Crocodiles are known to swallow stones to alter their buoyancy, so perhaps seals do also. Another possibility is that the presence of stones may reduce the pains of hunger when food is unavailable.

Once they have eaten, animals must digest and metabolize the food—that is, convert it into a usable form to nourish their bodies and provide them with energy. The thyroid gland is one of the principal regulators of the metabolism, and in cetaceans it is very large, probably because of the characteristically high metabolic rate of these animals.

Birds can usually be characterized by a gravel-filled crop and a high metabolic rate. Most seabirds, being fish-eaters, have relatively short intestines.

Morning takeoff. *A gull (above) quickly ascends after taking a meal from the surface of calm morning waters. Gravel in the bird's crop aids its digestion.*

Evening meal. *Illuminated by the setting sun, a sea lion (left) silently digests an evening meal, possibly aided by stones in his stomach.*

The Breath of Life

"Thar she blows," was the cry that rang out when the lookout aboard a whaling ship spotted the spouting of whales. The lookout and just about every other man on deck kept eyes peeled for the telltale blows of the whales, for that was the first and often the only obvious sign of the great beasts. With their blowholes and their dark coloration, the whales might otherwise have escaped the notice of the hunters.

The whalers could even tell what kind of whale they had spotted by the character of the blow, because the structure of the blowhole gives the spout a characteristic shape. Right whales, for example, have a double, V-shaped blow. The blue whale has a blow that is shaped like an inverted pear and is quite high, while the blows of finbacks and sei whales have the same shape but are smaller spouts than those of their blue relatives. The sperm whale's blow is pear-shaped but angled 45° forward and to the left because of the blowhole's location on the left side of the head; it is quite long in duration. A blowing whale makes a low moaning sound as it exhales, followed by a softer sucking sound as it inhales before snapping its blowhole shut and submerging. Harbor porpoises, no more than five feet long, sigh as they exhale and sip in a quick breath, then disappear beneath the waves.

As near as anyone can determine, the visible portion of the blow is a combination of two factors. One is the animal's exhalation: warm and humid from its time in the lungs, the air condenses as it strikes the cooler outside atmosphere. The other, according to a pair of

The **right whales** *have a characteristic V-shaped blow that is unique. Whales exchange as much as 95 percent of the air in their lungs with each breath they take, compared to 15 percent for man.*

RESPIRATORY RATE AT REST (Breaths Per Minute)

0　1　2　3　4　5　　　　　10　　　　　15

CALIFORNIA SEA LION
6

MAN
15

BOTTLENOSE DOLPHIN
3-4

ORCA
.8

British zoologists, is a foamy substance normally found in the cetaceans' trachea that is forced out in particles with the exhalation.

The relative size of the lungs of cetaceans and terrestrial animals varies, but what is more significant is the great difference in the amount of lung space that is used each time the animal breathes. Cetaceans exchange between 80 and 95 percent of the air in their lungs with each breath they draw, while man exchanges only 15 to 25 percent, pinnipeds around 35 percent, and sirenians, which are second only to the cetaceans in their adaptation to aquatic life, about 50 percent.

Other characteristics in the respiratory system of cetaceans include the exceptional development of diaphragm muscles and the far greater number of floating ribs (those attached only at the backbone and free of the breastbone on the ventral side or those

Slow breathing in the sea. The rate of breathing at rest is dependent on both body size and efficiency of oxygen exchange. Usually the bigger the animal, the less often it must breath.

not attached to either breastbone or backbone). Among dolphins there are at least four pairs of floating ribs, and among the right whales only one pair are not floating. Man has only two pair of ribs attached only at one end. These characteristics give the cetaceans a more flexible and more powerful breathing pump, explaining in part how they can exchange more than 80 percent of the air in the lungs each time they breathe. It explains also why it takes a rorqual only two seconds to exhale and inhale 1500 gallons of air, while it takes man four seconds to exhale and inhale a pint of air. The flexible chest allows for normal and harmless pressure squeeze on the whale as it descends to pressures of hundreds of pounds per square inch.

57

DURATION OF DIVES
MINUTES

PUFFIN .5
MAN 3.5
SEA OTTER 4.5
SEA LION 15
SPERM WHALE 75

How Deep? How Long?

On August 12, 1968, Robert Croft, a U.S. Navy diving instructor, took a deep breath and plunged into the deep blue waters of the Gulf Stream off Fort Lauderdale, Florida. He went to a depth of 240 feet, where three underwater cinematographers, who were filming this record dive, observed him. At that depth he stopped, turned, and headed for the surface. Two minutes and 28 seconds after he began his descent, he broke the surface, having dived deeper on a single held breath than any other human being. The underwater cameramen later were able to report to the medical researchers that Croft's chest was squeezed noticeably smaller by the 108 pounds per square inch (psi) pressure on him and that his abdomen too had a caved-in appearance. Normal atmospheric pressure at sea level is 14.8 psi.

Croft is an exceptional man in several ways. First, his physical condition, developed by daily held-breath dives to 100 feet, is excellent. He also has an enormous chest and lung capacity, which is at least partly the result of having rickets as a child. The effect of the soft-bone disease was to deform his chest and give him that tremendous capacity.

By contrast, most breath-holding divers in good physical condition can expect to reach a depth of about 75 feet in perhaps a minute and a half of underwater time.

For duration of underwater time, a number of "show divers" of the 1880s and 1890s made dives in glass-walled tanks before crowds of spectators for four and a half to four and three-quarter minutes.

Holding your breath. *The sperm whale is such a fine performer that the maximum duration of its dives must be put on a separate scale. Man can at least outclass the puffin in breath-holding.*

Cruising along the surface of the sea, some whale species may breathe once every two or three minutes, although after a particularly lengthy dive they may "pant" by breathing five or six times in a minute, until they get their breath back. At the other end of the scale, the little piked whale, a rorqual, has been observed to remain underwater for two full hours. Sperm whales, who usually remain down 50 minutes per dive, have been clocked at 90 minutes many times. Right whales commonly remain submerged for an hour at a time, while the piked whale's bigger relatives among the rorquals usually dive for 40 minutes and cruise the surface for about 10 minutes between dives. Dolphins and porpoises are far less spectacular, remaining down five minutes during their normally shallow dives. They seldom exceed 15 minutes. They usually sandwich several minutes of rest at two respirations a minute between dives. In emergencies any of these cetaceans may remain submerged below for longer periods of time.

Seals and sea lions, surprisingly, often remain underwater longer than dolphins and porpoises. Their dives are frequently as long as 15 minutes in duration. Steller's sea cow had been clocked at 16 minutes, and a Weddell seal made a controlled dive that lasted 45 minutes. The sea otter, less adapted to life in and under water, manages five-minute dives, while the polar bear, least adapted to marine life of the animals mentioned here and probably more properly classed as a terrestrial animal that lives by the sea, can remain underwater on a single breath for about a minute and a half.

Deep diving mammals. *This chart too must be adjusted for the extreme diving abilities of the Weddell seal and sperm whale. Unaided, man can barely penetrate the surface of the sea.*

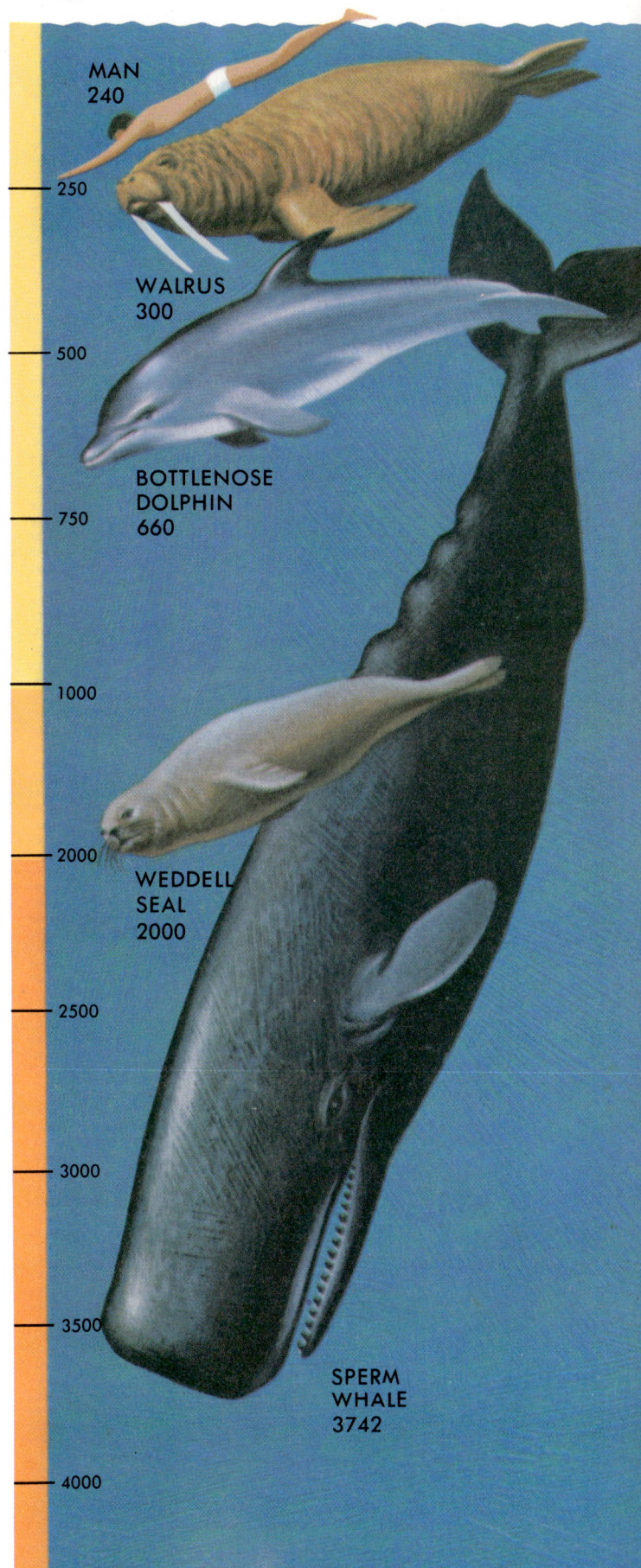

RECORDED MAXIMUM DEPTH OF DIVES
FEET

MAN 240
WALRUS 300
BOTTLENOSE DOLPHIN 660
WEDDELL SEAL 2000
SPERM WHALE 3742

BRADYCARDIA • Heart Rate Before and During Dive

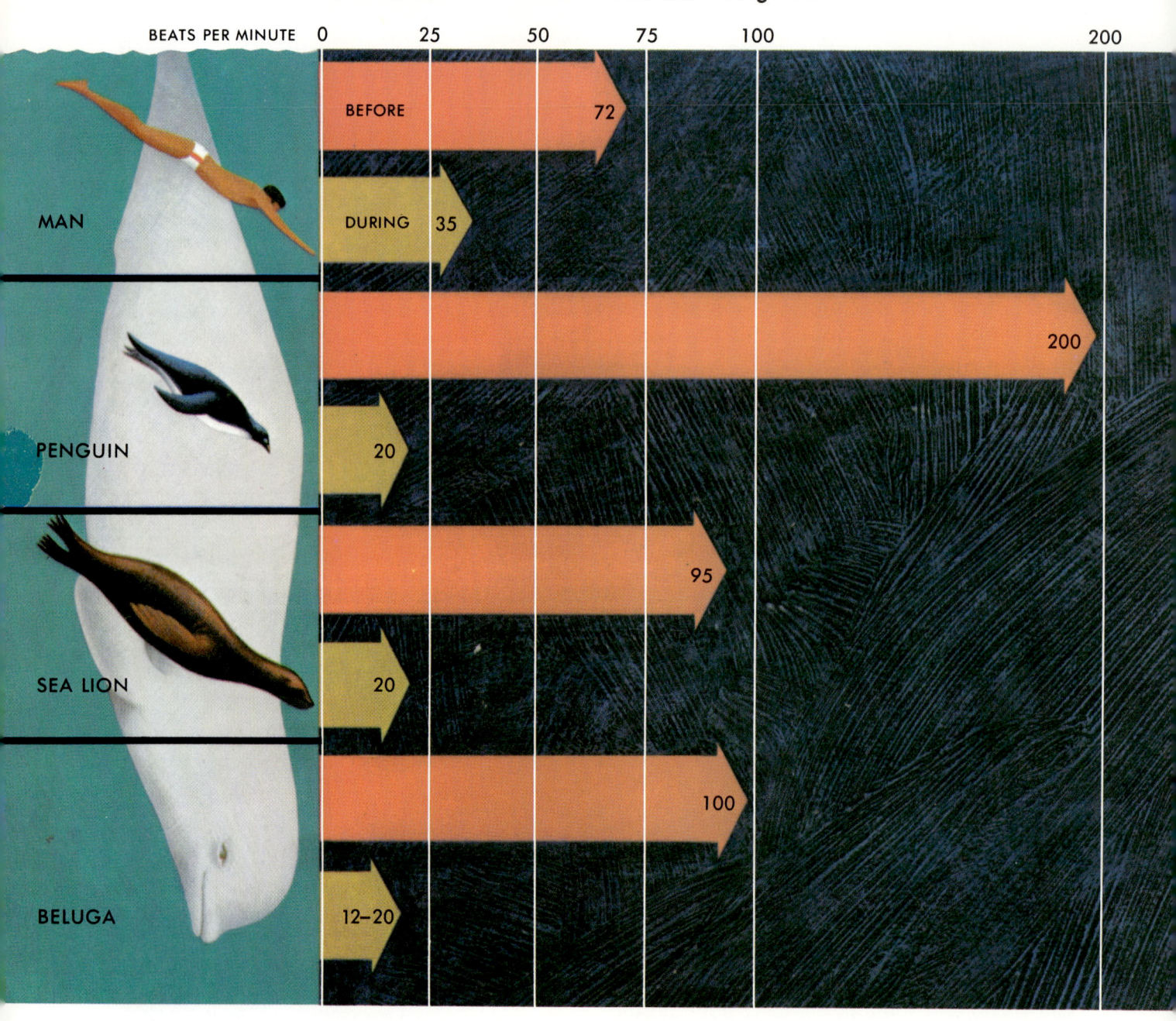

	BEATS PER MINUTE	
MAN	BEFORE: 72	DURING: 35
PENGUIN	BEFORE: 200	DURING: 20
SEA LION	BEFORE: 95	DURING: 20
BELUGA	BEFORE: 100	DURING: 12–20

Adaptations for Diving

When a new cross-harbor traffic tunnel was being built in Boston a number of years ago, high absenteeism among the sandhogs digging it cost the contractor cash penalties for falling behind schedule. A doctor at a city hospital who happened also to be a diver quickly comprehended the problem and how to correct it. Dr. Donald Butterfield knew the diggers worked in an atmosphere of compressed air that kept the mud and water of Boston Harbor from their excavations. They worked in pressures equal to three times atmospheric pressure; they returned to the lower pressure of the atmosphere when they finished work for the day. Many of the men became sick with pain and nausea. Dr. Butterfield ascertained they were suffering the well-known decompression sickness. He

Bradycardia means the slowdown of heart action, and for diving mammals it means much more time to dive. The rate of slowing of man's heartbeat in water is a fraction of that in diving mammals.

scheduled slow decompression for the diggers at the end of each workday and enforced the schedule. The drop in absenteeism was dramatic, work speeded up, and the project was finished on time.

Aquatic animals similarly experience increased pressure when they dive, but these diving animals never suffer decompression sicknesses when they return to the surface and to lower pressures. However, the circumstances are fundamentally different. Physiologists reason that diving animals come back to the surface for each breath and thus carry only a fixed amount of air with a limited amount of nitrogen in it. Caisson workers or divers, on the other hand, are breathing a constantly renewed supply of compressed air. Nitrogen dissolves in their blood on descending to high pressures. On ascending quickly to lower pressures, the nitrogen comes out of solution in bubbles much the same as carbon dioxide comes out of solution when a bottle of champagne is opened. The nitrogen bubbles often collect in joints and areas of old bone fractures, causing pain and crippling, a condition called caisson disease, or the bends. When the bubbles coalesce into large gas pockets and block the flow of blood, sections of the body may become asphyxiated. When the brain or nervous tissues are concerned, severe paralysis follows, and if lungs and heart are choked, the patient dies.

More answers are still needed. How, scientists want to know, can these marine animals dive and dive again? They must accumulate quantities of nitrogen theoretically sufficient to cause the bends in the long run. And, in the first place, how can seals, dolphins and whales remain underwater for such long periods without fresh air? There is no simple answer to such questions, but we know now that many physiological reactions in the respiratory and circulatory systems combine and make such performances possible. Research has shown that when any air-breathing animal dives, or even puts its face in the water its heart beats slower. As the heart beats slower, the blood supply to less essential areas of the body is shut off by the sphincter muscles of some arteries. The heart and brain and a few other important organs such as the liver, which require a constant supply of oxygenated blood, remain irrigated; other organs, like the kidneys, as well as the extremities, are shut off. Digestion ceases. Thus less oxygen is needed during the dive. Now this very oxygen comes from reserves that are much more important than in other animals. But this oxygen is not stored in a gaseous state; the lungs of marine mammals are rather small, and in many cases they are either emptied before a dive, or they are compressed by the surrounding water pressure during dives to such a degree that the remaining volume of air is isolated from the active part of the lungs. Oxygen, in fact, is accumulated either in chemical combination as oxides in blood and muscles or dissolved in organic liquids and tissues.

*Unique physiological adaptations give aquatic animals, like these **dolphins,** the ability to stay underwater for long periods of time before they have to return to the surface to breathe.*

*This **right whale**, blowing and gathering air, undergoes slowing of its heartbeat when it dives. This and its ability to cut off blood flow enables it to dive to relatively great depths for extended periods.*

Modifications of Circulation and Respiration

All marine mammals have a very large proportion of blood in their bodies, and their blood is very rich in hemoglobin. Their muscles are also much richer in myoglobin than those of land mammals and man. Accordingly, at rest, they can store large quantities of oxygen.

However, as we have seen earlier, during dives this muscular oxygen supply is short of what is needed for active muscles that have been shut off from blood circulation. In fact, waste materials and lactic acid generated during muscular work accumulate in the muscles during a dive, and it is only upon returning to the surface and resumation of respiration and normal blood circulation that these wastes are washed away and burned by fresh blood rich in oxygen.

Moreover, diving animals tolerate in their circulatory system a much higher rate of CO_2 than other animals before they feel the need to breathe, and this enables them to use the oxygen stocked in their blood more efficiently.

Yet, at the end of a long dive, a whale comes back to the surface "panting" and stays afloat to ventilate its lungs, drawing a series of deep breaths. The whale's lungs have a greater efficiency than those of man, the higher quality of its blood permits a faster elimination of CO_2 and the storage of more oxygen, and the whale quickly gets rid of the dive's waste materials. In a second phase it replenishes its oxygen supply at a slower pace, depending upon the species, of the order of one respiration per minute for about ten minutes. Then the whale is ready for another deep, long, active dive. These physiological necessities are well known by whalers who chase the whale without harpooning it until it is exhausted, coming back more often to the surface, and much more vulnerable.

The same basic improvements of the essential physiological processes occur at various degrees with all diving mammals and birds. It is necessary, however, to insist that these mechanisms are only improved; they exist fundamentally to a lesser extent in all birds and mammals. Physiological adaptations to diving are also present in the marine iguana which goes to sea for food.

A slowdown of the heart-rhythm also occurs in human divers, though physiologists have not yet determined what causes such a reflex; with intensive training the slowdown increases. With training a diver can match the Japanese amas who are born of generations of professional diving women. But such progress is very small compared with that of the Weddell seal or the marine iguana which are both capable of spectacular alterations of cardiac rhythm and of shutting off blood circulation to large sections of their bodies. It is probably these circulatory modifications, even more than the oxygen storage capacity, that are responsible for the Weddell seal's ability to reach a depth of 2000 feet and to stay 45 minutes underwater.

*In **diving mammals** oxygen is stored in muscle tissue. These animals can therefore remain underwater longer than their terrestrial counterparts like man. Here a harbor seal swims near bottom.*

Birth at Sea

When a human is born, it usually emerges from the birth canal headfirst, face up, with the mother being assisted by another person. In cetaceans, however, calves are born underwater, tail first and with no outside assistance, although sometimes another adult female stands by the mother who is delivering, ready to help in any way it can. As soon as the calf's head is clear, the mother moves away, and the umbilical cord breaks close to the baby. The calf, able to swim as soon as it is born, swims or is pushed by its mother or the midwife dolphin to the surface. There the air on its skin stimulates the baby to take its first breath, much like a doctor's slap on a baby's buttocks often causes it to draw in its first air. Sometimes the new dolphin's umbilical cord does not break, and the calf, swimming away, pulls the placenta out of its mother's womb and is dragged down to die from lack of air. Cetaceans have never been known to bite the umbilical cord as many other animals do, nor to eat the placenta, which is passed several hours after birth. In reality what little we know is based upon observing dolphins in captivity and what occurs with whales is largely unknown. Most current knowledge is based on births of captive dolphins or on rorquals killed while in labor when they are easy marks for the harpoon gunners. It is known that cetaceans bear only one calf and exceptionally twins after a gestation period of 9 to 14 months, depending on the species. A harbor porpoise may weigh 25 pounds at birth, while a blue whale with a gestation period only a few weeks longer may weigh 6 or 7 tons.

The mother suckles her young from seven months in the case of a finback whale to 18 months in other species. She has two mammary glands enclosed in folds on either side of her genital fold. She forces her milk out by muscular contractions, and the calf gets the rich milk at the opening of the mammary fold. The fat content of cetaceans' milk is extraordinarily high, and the water content very low when compared to Grade A cow's milk. Harbor porpoise milk is close to 46 percent fat and the blue whale's is 42 percent fat, against 4 percent for cow's milk. While the porpoise's and the blue whale's milk are a little more than 40 percent water, cow's milk is 87.5 percent water. Cetacean milk is also much higher in protein and other solids.

The age at which cetaceans reach sexual maturity varies with the species, and it is open to some speculation. Bottle-nosed dolphins are sexually mature at three or four years, but are able to copulate within a few months after birth. These are the cetaceans best known to man because of the large numbers in captivity that can be studied.

Courtship among whales, porpoises, and dolphins is a lengthy process, lasting many hours and building in intensity as the paired animals swim close together, rubbing and biting

First breath. Dolphins are born tailfirst, for soon after they emerge they must take their first breath of air. Dolphin calves swim or are helped to the surface, where contact with air triggers breathing.

each other more frequently until a high-pitched excitement climaxes with copulation for a matter of seconds. A fold within the female's genitals apparently retains the sperm and keeps water, which would kill the sperm, from entering the female's genitals. But for most cetaceans we know little of their behavior in the wild.

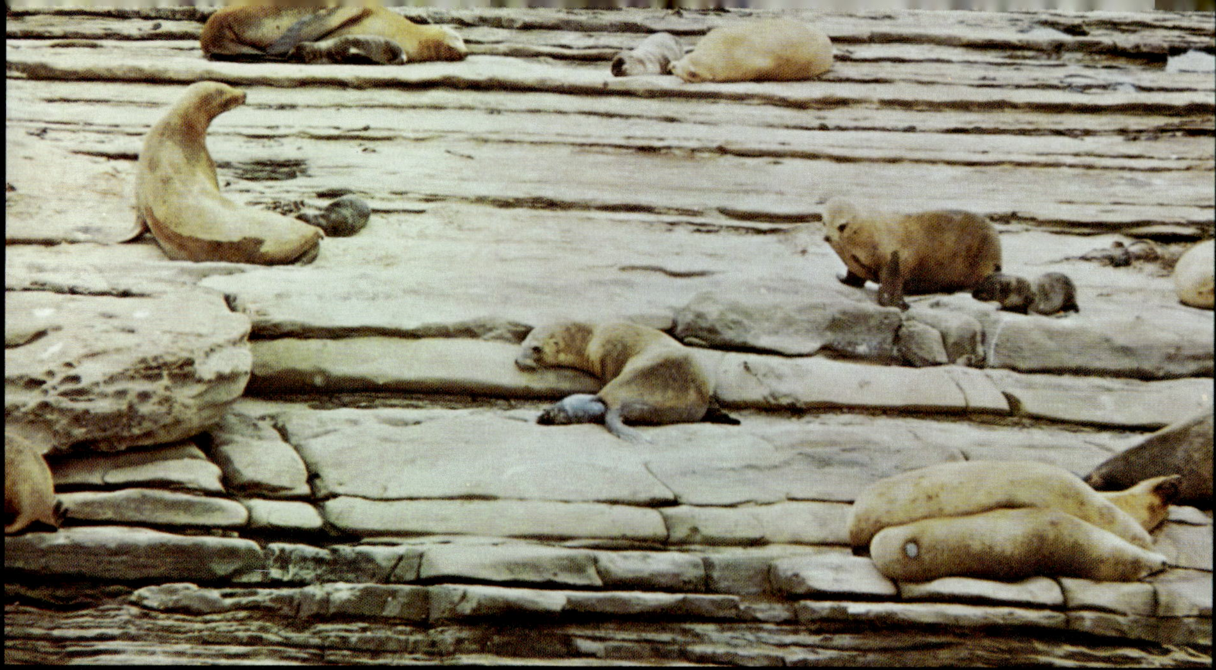
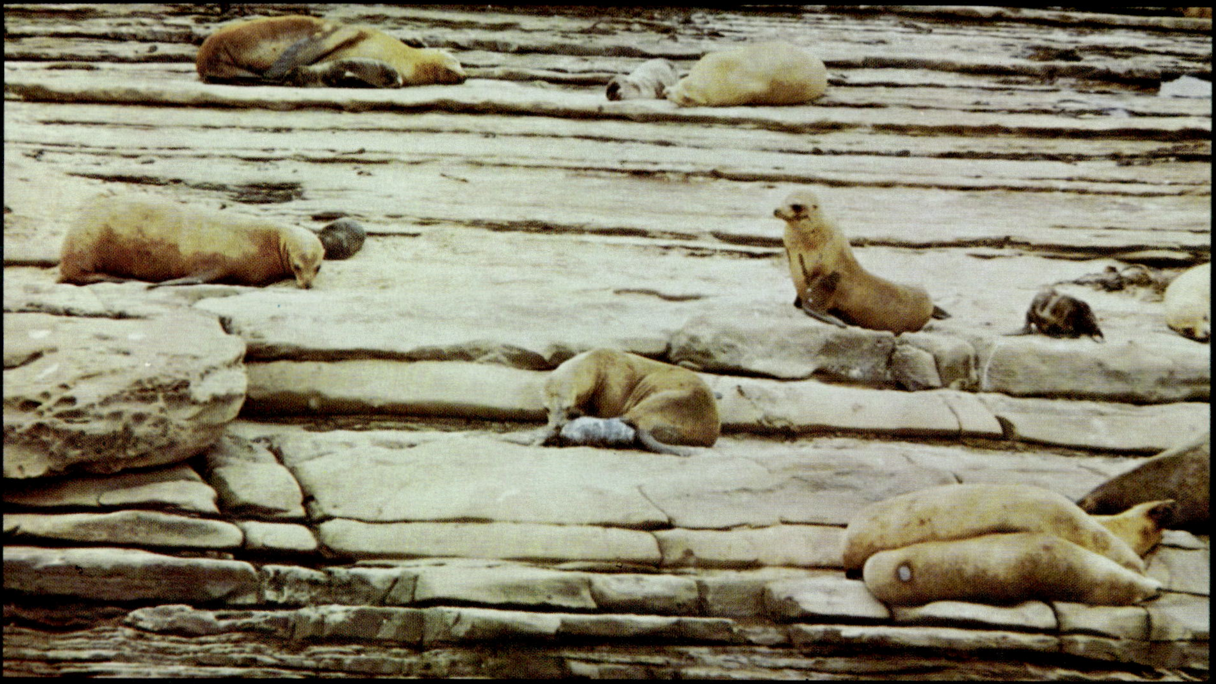
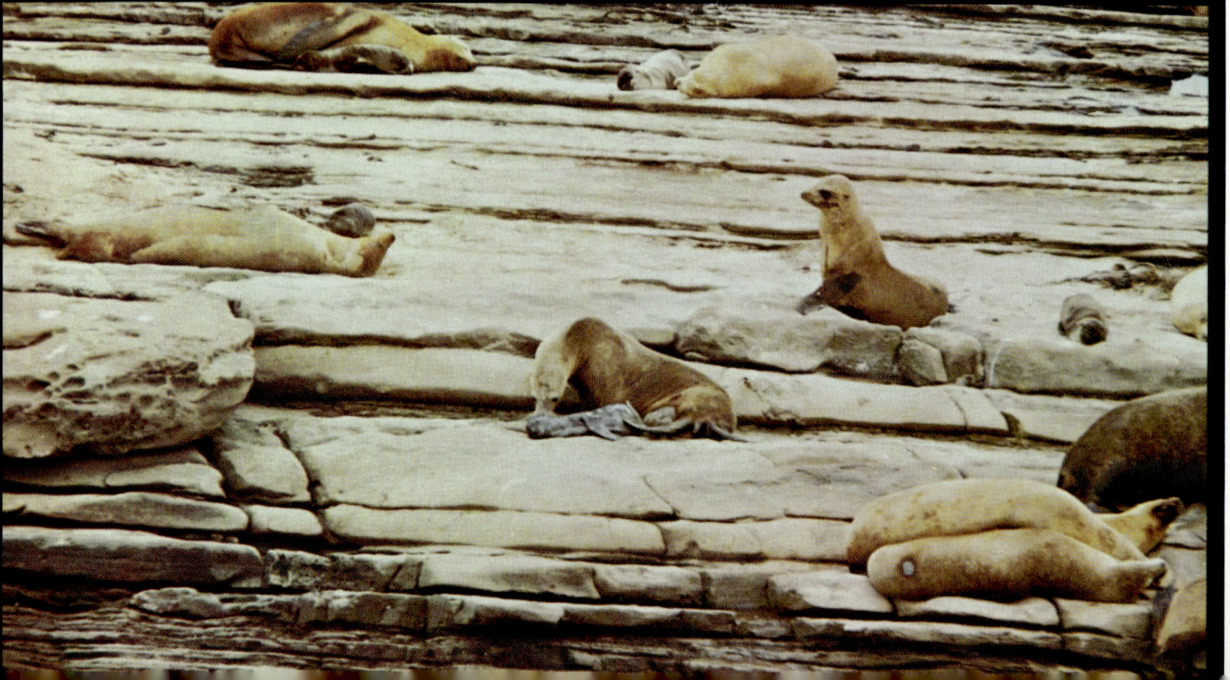

Birth Ashore

In most species a bull seal becomes sexually mature when it is four to six years old. It does not achieve a harem of its own, however, until it is 10 or 12 years old. Challenges are common but meet with little success. The delay assures that only the strongest males, those able to defeat other bulls to establish territories on the breeding grounds, will sire the next generation.

Typically, the bulls come ashore and battle with each other for territories, then seek to attract as many females to their domain as they can hold when the cows arrive several weeks after them. The harems vary in size with the species. Elephant seals commonly have 20 to 40 cows, although some of 100 have been counted. The fur seals of the Pribilof Islands may hold as many as 50. Steller's sea lions usually have 10 to 15.

When the females arrive, they give birth to the pups conceived the previous year and begin nursing them. The bulls, meanwhile, fight among themselves to keep the cows in their territories or to lure cows from the territories of others. The females in many seal species are completely indifferent about which bulls they mate with. Those at the edge of one bull's area may wander over and mate with two or more neighboring bulls. In this confusion, many newborn calves get crushed by the excited males. A few weeks after giving birth, the cows are ready to mate with whichever male happens along. Occasionally one of the younger bulls with no harem of his own invades a beachmaster's harem and mates with one or more cows there. The beachmaster usually chases these younger bulls away before they can get to any of the cows, but if the beachmaster has been challenged on one front of its territory, another bull may sneak in on another front.

From two to four months after mating, implantation takes place. In most other animals, implantation takes place as soon as the sperm penetrates the ovum. This delayed implantation in seals permits yearly breeding and births in animals living apart the rest of the year. Climate, hormonal factors, and availability may also be related to delayed implantation. The exception is the walrus in which implantation takes place just after copulation, and gestation is about a year.

One and rarely two pups are born to pinniped mothers. They nurse their young from any of their four mammary glands on milk that, like cetaceans' milk, is high in fat content—more than 50 percent in some species.

Such rich milk causes pups to grow rapidly. Weddell seals can gain about seven pounds a day while southern elephant seal pups add as much as fifteen pounds a day from feeding on their mother's milk.

Giving birth. The startling sequence (opposite) shows a California sea lion giving birth. Pups are born in June each year and nursed by their mothers for five to six months.

A Galápagos sea lion cares for her newly born pup. Pups are nurtured on a rich milk provided by mother. It is very rare for a pinniped to give birth to more than one pup per year.

Breeding Birds

The story of procreation for seabirds also takes place ashore, but it takes on a very different pattern. Breeding and nesting season in the spring in temperate zones is the only time some birds ever come ashore. Many other species that do roost ashore normally spend much more time away from the water during nesting season, returning to it only to feed and to gather food for the nestlings. Among birds fertilization is internal as it is among mammals. The eggs develop inside the female until they are mature. Then she lays them. Depending on the species, one or both partners build the nest, incubate the eggs, and raise the young. In some species one parent stays at the nest, while the other gathers food. In others both parents may leave the nest for brief periods. A number of bird species capture food, swallow it, then regurgitate it predigested for their young. The young grow, mature, and eventually learn to fly. As soon as they can fend for themselves, these far-ranging birds head to the open sea.

Courtship behavior of birds is often colorful as it is in the case of the frigate bird. Frigate birds have long, slender beaks, which are sharply hooked at the end. Between the edges of the lower mandible on the chin there is a patch of naked skin. To attract a female in breeding season, the male frigate inflates this bright red patch like a shiny scarlet balloon in a spectacular nuptial display. Colonies of breeding frigates build clumsy platform nests on tropical islands in trees or bushes or sometimes on rocks. These nests must be guarded constantly against other marauding frigate birds which would steal the construction material, the single egg, or the newly hatched chick.

Sexual maturity comes early to some birds, but in many species mating and nesting is only accomplished at the age of six or seven years, after several unsuccessful attempts. Such is the case of the albatross, which does not breed until it is at least six years old. These birds are faithful to their nesting sites, returning to them after their world-ranging voyages. During courtship, males dance around the females, and mating pairs dance with each other, wings outspread, to the sounds of groans and snapping bills. The female lays a single egg in a nest built of mud, and both parents guard the chick for some weeks. The young albatross is dependent on its parents for food for as long as eight or nine months in some species, so the albatross breed only in alternate years.

Australian masked gannet (left). The nest of this bird is little more than a pad of twigs on ground.

Pacific kittiwakes (below) nest and raise their young in inaccessible ledges that border the sea.

Chapter IV. The Senses

"His eye was as big as both my fists and blue as a baby's. When he looked at us with this bleary blue eye, I could see malice in it, but that may have been because his size was so imposing or that he knew we would do him in. He had startled us, breaking water next to our boat as he did, because Chalmers on the middle oar had thumped about so much and caused the mate to curse him violently." This excerpt from the unpublished letter of a seaman aboard a whaling vessel in 1853 must, of course, be stripped of the inevitable fabrication that has always been and still is a trait

> **"Mammals and sea birds have senses similar to ours, but modified to function in water."**

of whalers. *Calypso* divers have many times been eye-to-eye, a few feet apart, with all sorts of whales, and in no case have we noticed any "malice" in their look. Nevertheless, the letter tells us that even the nineteenth-century whaling men knew that whales had senses much as humans do.

The mammals and birds of the sea do have much the same senses we have, but they have been modified to function in the water. Water, no matter how clear, is never as transparent as air, and its transparency varies considerably from one location to another far more dramatically than air's. Very clear water permits visibility of little more than a hundred feet. Particles of matter—like minute plants and animals—reduce visibility in water, reflecting and scattering light, and absorbing the colors of the spectrum one at a time. The eyes of aquatic animals cope more or less with these many variables. The same holds true for sound in the water. The density of water is 800 times greater than that of air, and molecules of water are much closer to each other than are air molecules to each other. The water is far better as a transmitter of sound than air, and sound travels much faster in water than in air. Sound waves are scattered or reflected by objects. The hearing organs of animals in the sea take advantage of these physical facts. Taste and touch, being senses dependent on the closest possible contact, are least affected by the marine environment, because there is less opportunity for the water to affect the stimulation of such senses.

Fish and those invertebrates equipped to extract dissolved oxygen directly from the waters they live in have accordingly various sensitive mucuses bathed by water and often well-developed senses of smell. But warm-blooded animals that must rise to the surface to take air can use a sense of smell only occasionally and only when above the sea's surface. To replace the ability to sense distant stimuli, air-breathing creatures of the aquatic world have been able to use their other senses as most sensitive and efficient underwater receptors. They have developed a highly discriminating echolocation system which might pass for a new sense related to hearing in the case of cetaceans and pinnipeds, though this is simply an adaptation of organs and functions that most terrestrial animals have but use differently. Interestingly, the airborne bat, which lives in a three-dimensional world like the cetaceans and pinnipeds, is the only other mammal that has developed a comparable acoustical sense, and it has done so for the same purpose.

*A **California gray whale** spyhops for a look around the coastal waters. The grays often poke their heads above water while migrating along the California coast to or from breeding grounds.*

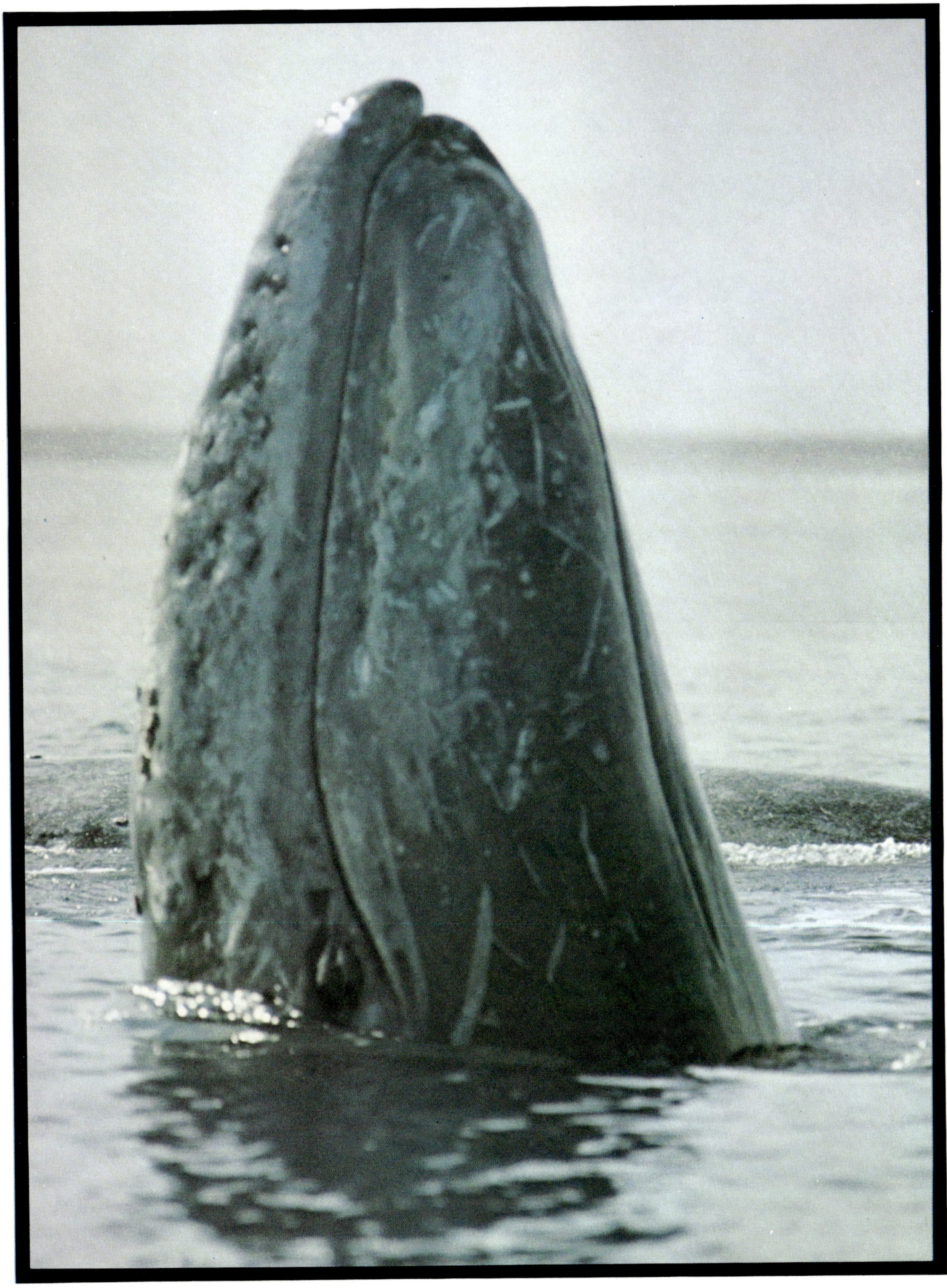

Sight in the Sea

Dolphins that have been blindfolded experimentally can navigate without difficulty. However, cetaceans have good eyesight. Vision is an important sense to them and to pinnipeds, sirenians, and the various species of birds that dive underwater.

Man can record and study the sounds uttered by animals, and thus we can have a fairly good idea of what their ears transmit

Pinniped eyes are well adapted to sight in the sea. Some deep diving species have much the same mechanisms as nocturnal land animals for seeing where little light penetrates. Their lens is curved, allowing an underwater image to be focused on the retina.

to their brains, in other words what the acoustical image of sound is. On the contrary, the optical image formed by a complex eye like that of an insect, for example, is decoded in the brain and is not directly accessible to the investigator. The animals are unable to communicate such images to us and the real optical image formed on the retina of a *living* eye is only a fugitive shadow, very difficult to reconstitute when one experiments with a dead and deformed eye. The study of vision is extremely difficult and must take into consideration at least as much behavioral observation as physiological studies.

In Volume IV *(Window in the Sea)* we saw that light passing through water is scattered and absorbed so that vision becomes limited by the availability of light. Also that water's refractive index is different from air and comparable to that of glass or of lucite. Thus the eyes of the creatures that must see beneath the surface display drastic modifications suitable to the differing conditions.

When light rays from the objects in sight enter the eye, they are deflected by crossing media each of which have a specific refracting index through curved surfaces. For man in air these media are the cornea, the vitreous humour, and the crystalline lens. All help form a sharp image on the retina.

For a whale in the sea the refracting indexes of the cornea and the humour are very much the same as the index of seawater. The crystalline lens is the only element of the eye that is operational in building up images. That lens is much more spherical than the lens of man and thus more convergent. But it seems unable to accommodate and probably has almost a fixed focus.

Another difference is that the eye as a whole is not spherical, but strongly flattened. The surface of entry of the cornea is almost flat, thus there must be very little distortion when the eye is in the air. This is in accordance with the behavioral observations that cetaceans have good vision in air and water.

*The eye of a **cormorant** is not spherical, but flat. This gives the bird excellent vision in air as well as underwater, allowing it to sight fish from above and dive underwater to catch them.*

Amphibious Eyes

We have seen that the eyes of marine mammals are modified, without changing their basic anatomy, to be able to see almost as well in air as underwater.

The changes in hydrostatic pressure these animals experience during deep dives cannot have any more influence on the eyes than on other organs. The pressure equilibrium is exact between incompressible body fluids and tissues, the surrounding water, and the blood that irrigates all tissues.

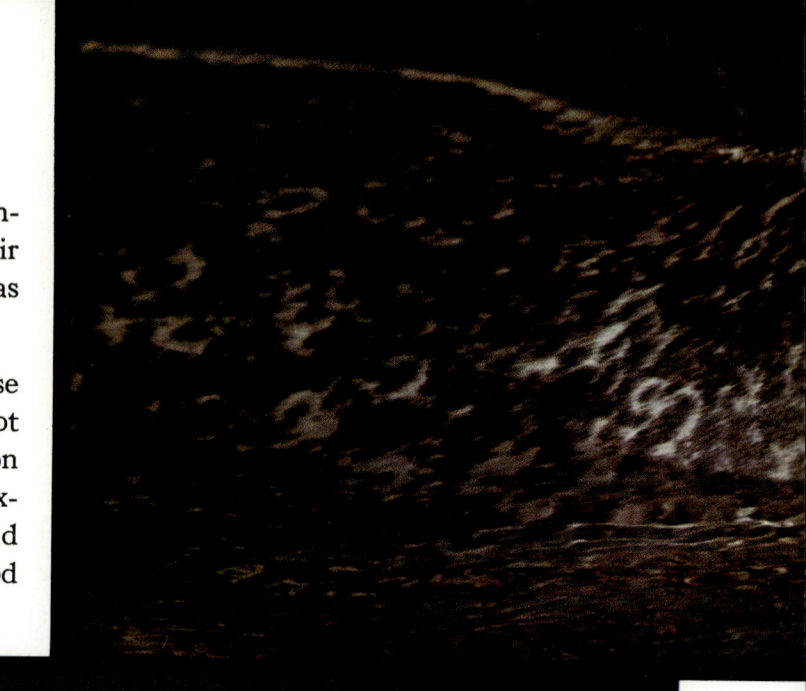

good vision these animals have. If a powerful attacking male crushes a baby, it is not due to myopia but rather to a proud indifference.

Cetaceans do not really need to see well in air. But the "spying" behavior of sperm and gray whales that emerge vertically to observe what is happening around them is indicative of their amphibious sight. The orca occasionally attacks seals that are resting on ice floes, using such ability. In marinelands, bottle-nosed dolphins and pilot whales perform tricks like throwing objects in the air, and these require excellent vision in air.

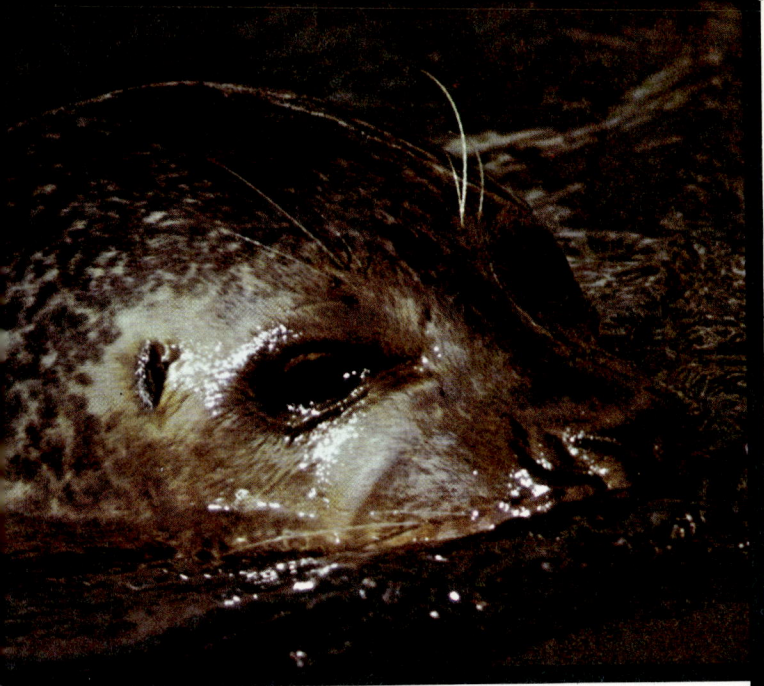

Siberian harbor seal (above) can see well through elliptical pupils, underwater and above surface.
Galápagos sea lions (left), like all pinnipeds, have a third eyelid for better underwater vision.
The **bottle-nosed dolphin** (right), like all cetaceans, sheds no tears. Its vision is relatively good.

Whales shed no tears, but seals cry constantly when they are on the beach. The whales spend all their lives in water and have no need for salty tears to cleanse their eyes. In place of tears they have an oily secretion that protects their eyes from irritating particles in suspension in the water. Pinnipeds, on the other hand, spend long periods out of water, often on dry and dusty beaches. Because they have no tear ducts, tears flow constantly from their tear glands.

Pinnipeds have another protection that serves them underwater. A third transparent eyelid, or nictitating membrane, slides across the eye to protect against injuries. In some cases the nictitating membrane has optical properties that help correct the observations due to the fact that the surface of entry is not absolutely flat.

The elephant seal clearly sees its rival when it is fiercely fighting and injuring it with rapid strokes. The fact that blows are often skillfully avoided is another proof of the

For most diving birds the problem is quite different. They spot their prey from above, through the surface, and must take into consideration the fact that the fish or squid appear to be where they are not, and correct for that before they attack. Once under the surface, they make small adjustments that do not require very sharp vision.

Nevertheless, some birds, like cormorants and puffins, swim long distances under the sea to hunt their prey and must have some way to adjust their eyes.

Hearing and Acoustical Sense

When nineteenth-century whalers launched their long boats from their ship, they took great care not to bump the boats against the ship's hull as they lowered away. Once they were in the water, stroking toward their intended victims, oarsmen were constantly being cautioned in hushed tones by the mate not to splash as they rowed. Superstition was not the reason for the great care taken to avoid noise. The whalers knew that cetaceans, having excellent hearing, would be alerted and possibly frightened off by unexpected sounds like the thump of an oar against the side of a boat. It is quite the opposite in the Solomon Islands where the natives hunt dolphins and porpoises by using noise. Banging rocks together underwater frightens the porpoises and drives them toward shore where hunters await them. Today's whalers bypass this hearing acuity in using "supercatchers," very fast chasing boats. They pursue the animals until they are out of breath.

Cetaceans' ears are similar to those of terrestrial animals in several respects. There are, however, notable differences, which enable them to hear better underwater. They have no external ears nor do the true seals and walruses. Sea lions have small ears that lie flat against the head so streamlining is not disrupted. Birds have only a small opening covered with feathers.

Recent studies on two species of dolphins have shown that their mechanisms of hearing are very different from most other mammals. Some of the differences include: the outer ear canal is filled with a plug, surgical removal of the eardrum does not impair hearing, and the auditory ossicles, which normally transmit sound from the eardrum to the inner ear, are not even connected to the eardrum.

In addition researchers have found that high frequency sound may be initially received in the lower jaw. By directing a narrow beam of sound at the head of the dolphins it was found that the skull was acoustically isolated from the inner ear but that the sides of the lower jaw were very sensitive and effectively transmitted sound to the ear. It was concluded that dolphins may hear by translational bone conduction. The frequency of sound used in these studies was 20,000 Herz (cycles per second), although some dolphins can hear as high as 200,000 Herz. For lower frequency sounds in the human hearing range, sound probably reaches the inner ear through the body tissues and bones in general.

Seals were thought at one time to be deaf, or nearly so, because men could walk up to them in the midst of their colonies without attracting attention. Noise in those colonies, however, is so great and constant, that the seals probably were unable to differentiate between the human sounds and those of the rookery. In air, harbor seals may respond to sounds of up to 160,000 Herz, while underwater they can discern sounds up to 12,000 Herz. Experiments show good directional hearing in many pinnipeds.

A characteristic voice. *Many whales can identify the sounds of an orca and avoid it.*

Sounds of the Deep

The songs of birds have often inspired composers, but not until recently did the song of a whale inspire them. The composer Alan Hovhaness, however, has written music based on the eery, pleading, groaning, sighing calls of the humpback whales. To many, the songs of these whales have a sad sound of great strength confined, almost a pleading sound, begging some other whale to answer. To all who have heard recordings of the humpbacks, the sounds are very moving.

Humpbacks whales, however, are not the only ones that produce sounds. All cetaceans make sounds and use them in several ways. One use of phonations is for echolocation. The whale, porpoise, or dolphin sends a very high-frequency sound through the water and listens for the echo. Sounds are used also for communication among members of the same

The beluga whale can produce sound internally, under water, and out of water by the blow hole. These noises range from high squeaks and sharp whistles to scolding sounds.

species. When *Calypso* accidentally struck a baby sperm whale, it cried out with a shrill whistle that the crew could hear above water. In apparent response to its cries, adult whales from its pod came to see what was wrong and "chatted" among themselves, perhaps communicating simple ideas about the situation. Bottle-nosed dolphins' phonations have been studied more than those of any other cetaceans because captive bottle-noses are readily available at a score of oceanariums and aquariums. It was once thought that rorquals made no underwater sounds, but only because no one had heard one. A cetologist and an electronics expert at Woods Hole Oceanographic Institution, however, teamed to record a number of whales including blue and finback whales. William Schevill and William Watkins produced recordings of whales long before the whale sounds inspired composers and won wide popularity.

A number of the whales are known to dive to depths where the light level is very low. The sperm whale, the deepest diver of them all, reaches areas that are well beyond the farthest limit of light penetration, descending often beyond 2000 feet in search of squid and cuttlefish. But if there is no light, how does the sperm whale find its quarry? The answer is presumed: by echolocation.

Confusion in the echolocation systems among pilot whales and porpoises may account for beaching of some of these animals. In the town of Wellfleet on Cape Cod, Massachusetts, pilot whales, 30 to 50 at a time, have repeatedly come close to shore and have been beached when the tide ebbed. They then die of dehydration and collapse of their lungs under their weight. As yet, no one has been able to explain how it is that an entire pod of whales is similarly affected. Just south of the space center on Florida's Atlantic coast, a pod of pilot whales came ashore on one occasion. Coast Guardsmen labored to get the whales off the beach and headed out to sea. But the whales were of a different mind. They turned around and headed back to the beach. Some persons have suggested suicidal intent was behind these mass beachings.

How cetaceans produce phonations is still an unanswered question. They have no vocal chords, but they do have folds of tissue and chambers associated with their windpipes which may be capable of sound production.

Underwater sound emissions of the bottle-nosed dolphin often are of ultrahigh frequency, beyond the hearing of man. These phonations serve as a sonar system for the animal which perceives their echoes.

Sounding Off

The cacophony of seabirds hovering over a school of small fish to capture a meal is one of the sounds of sea life. The barking of seals sunning on rocks just above water line is another. If we lower hydrophones into the water, we may also hear the underwater sounds of pinnipeds. Seals and sea lions are a vocal group of animals, and the phonations of Steller's sea lion, for example, can be heard up to five miles in air.

Unlike cetaceans, pinnipeds have vocal chords. They produce sounds both to threaten and to attract. The dominant animal in a group is usually the most vocal, emitting a variety of sounds described as barks, squeaks, growls, belches, rasps, and roars, depending on the species. It seems as if the animals of a species can recognize some individual voices.

Brown pelican and California sea lion *(opposite, top) appear to be arguing over a rock.*

The Hawaiian monk seal *(opposite, bottom) uses its underwater sounds for echolocation.*

Australian fur seals *(above) produce sounds to threaten, attract, and identify individuals.*

A California sea lion *(left) barks at a pup and awaits its reply to be sure the pup belongs to it.*

Aside from the sounds used as threats and attraction, some of the pinnipeds use their vocalizations while underwater in the same manner as cetaceans. The melodious song of the arctic bearded seal, as well as the plaintive calls of the antarctic Weddell seal can be heard by divers or recorded several miles away. Echolocation by means of series of clicks is rare among them. Elephant seals have none. In other species it is so well developed that a blind seal can survive quite well, being able to capture its fish with its guidance system. Yet there seems to be evidence that pinnipeds are not as well endowed with echolocation systems as dolphins.

Seabirds produce calls for feeding, threatening, alarm, and courting. Vocalizations are made by forcing air through the syrinx, an organ comparable to the human larynx.

Although not much research has been devoted to the vocalizations of manatees and dugongs, we do know they produce a few sounds underwater. These sounds are described simply as squeaks. So far, there is no evidence they use squeaks for echolocation.

Touch, Taste, and Smell

If a piece of clear glass is placed over a hole, most animals will not walk across the glass, fearing they will fall over a precipice. If, however, they have "whiskers," or vibrissae, on their muzzles, like cats and rats, and they touch the glass with their vibrissae, they will walk on the glass. Pinnipeds, sirenians, and few cetaceans have vibrissae, which presumably are sensitive enough to serve as sensory receptors. Those of the walrus and the sea otter are particularly sensitive.

The sense of touch goes beyond just the vibrissae, however. The tactile sense, as humans know it, seems to be well developed in most pinnipeds, but reaction to touch varies broadly. Elephant seals seem indifferent to crowding that forces them into bodily contact with each other. The Laysan monk seal, on the other hand, rarely touches even its own congeners, including its pups. During courtship, female California sea lions rub the males, and courting dolphins swim together in frequent contact with each other until they are ready to copulate. Divers in Florida

A sea lion cow (left) *pats its pup with a flipper. Both touch and smell are used to identify pups.*

waters report that manatees sometimes approach them, apparently seeking the stimulation of being patted.

The other major senses include smell for discerning distant stimuli and taste for stimuli directly in contact. Cetaceans, living their entire lives in water, have virtually no sense of smell: being air breathers, they can only smell in the air, which is of little use to them. They have a small, underdeveloped olfactory center in the brain. To receive and sense stimuli from a distance, they use instead

Below, a **seal** *nurses her pup. Touch and taste play an important role in this feeding pattern.*

82

Seals (above) *have a full set of whiskers that aid them in detecting objects around them.*

their acoustical sense. Pinnipeds too have a reduced sense of smell, used mostly by the cows seeking their pups out of a crowd. Even then the cows depend first on hearing the cries of their young and use smell only as a final verification. But sea otters have such a sensitive sense of smell that, according to one nineteenth-century observer, they can smell smoke for a distance of five miles. The same observer reported sea otters would wait until several tides had washed a beach before landing where humans had walked.

Bottle-nosed dolphins (right) *touch and rub against each other as part of their courtship.*

Animals that swallow their food without chewing, as cetaceans and pinnipeds do, have little need for a sense of taste and consequently have little. Seals have shown they can discern the difference between beef and horsemeat, which they usually refuse, and fish, which they prefer.

Other senses of marine animals, such as perception of heat, pain, and water currents on the body, are not very well understood. It seems apparent that all perceive heat and cold, which trigger their natural defenses against extremes of temperatures.

Chapter V. Marine Biographies

There is no such thing as a "balance of nature," because such a term suggests a stable equilibrium. The environment is continually changing; it is influenced not only by natural forces, but also by human activity, often with unforeseen and detrimental effect. In order to control these changes or adapt to them, it is necessary to understand what has happened, and how and why it happened. It then becomes at least possible to put an end to destructive practices or devise new procedures, perhaps even to undo some of the damage. An important way of accumulating this kind of knowledge comes from the study of animal life in relation to our own.

By tracing the development of an animal from conception through its embryonic stages to birth, we can learn about the past history of a species. In that earliest phase of life, an animal undergoes the stages of

> **"An animals life history can tell us enough to speculate on the future of that species."**

development its ancestors passed through in evolving to its present form. Later on in life, as an animal grows and develops from the immature to the juvenile to the adult form, its behavior changes as much as its physical appearance.

From the life history of an animal we can learn enough about the past and present to discern signs about what the future might be for that species. It is also essential to compare the life histories of animals and to explain the differences. Our knowledge of the living beings on this planet is still so narrow, however, that we are often surprised by the unpredictable consequences of biological happenings. Thus, when the cattle egret moved without human help from its home habitat in Africa across the Atlantic to the South American coast, it was a surprise. It was the first record of a species extending its range from the Old World to the New World. But the cattle egrets' subsequent spread to North America came as no great surprise to zoogeographers who, having been caught unaware by the first migration, were now ready for this northward move.

Such changes are part of the continual mixing, stirring up, and reshuffling that takes place in nature. When an animal spreads its habitat, it has a direct bearing on other animals in that area. The species may compete with native populations for food or space, for example. If those who imported the first English sparrows and starlings to the United States had considered the life history of these birds, they might not have brought them. Their effects on native populations of birds is still being felt. Having no natural enemies when they arrived in the U.S. and being an aggressive species, they were able to drive from their territories many local birds. Similarly, the influx of herring gulls in some coastal regions has driven the smaller, less competitive laughing gulls out of parts of their range. And the larger great black-backed gulls of more northern temperate provinces have also moved southward.

What we can learn about a species may tell us when it should be protected from man. Knowing what we do, at last, about the life history of the whales, we are sure that all species must get some real protection soon.

Sea lions and cormorants share seaside rocks, but both groups remain separate from each other. Both species inhabit the same waters off Pacific coast.

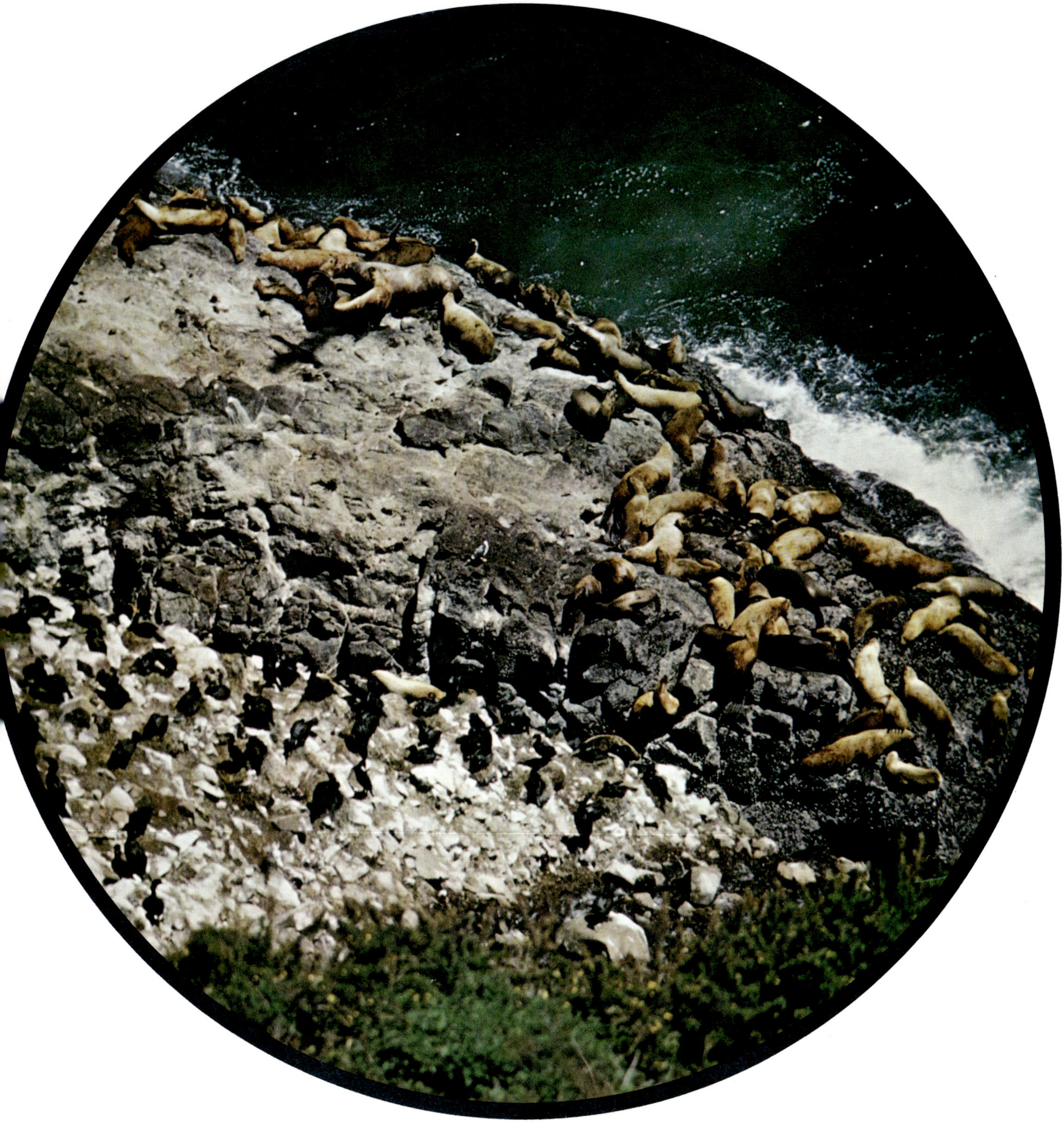

*This **horned puffin**, a rare West Coast relative of the Atlantic puffin, belongs to the alcid family, which includes the now-extinct great auk.*

A Sociable Seabird

The fat, little black-and-white puffin, with its brilliantly hued bill and its stubby wings, gives the impression of being a clown. But a more appropriate way of describing puffins is "sociable." After seven months at sea, the puffins come home to roost. Home is a burrow dug into the side of a shoreside cliff to a depth of more than three feet. Later in the season the puffins nesting in these burrows share them with razor-billed auks and black guillemots, to which they are related.

***Half a dozen herring** are held in careful order in the bill of one of these puffins. In winter plumage the bright colors of the bill fade considerably.*

The homecoming is a sudden mass arrival of the birds in the waters near their nesting grounds. For the first week the birds mill about among themselves, perhaps resting, perhaps picking out a likely partner. In the week or so before the puffins finally come ashore, they court and mate in the water.

The puffin is an affectionate bird. Once it has found a mate, it remains close to its partner. The male frequently touches and rubs the bill of its companion. For four or five weeks copulation takes place in the water repeatedly until the female goes ashore and lays a single egg in the burrow the pair has dug.

From late March or early April, when the birds first pair off, they socialize little with other birds around them, but after mating has begun and burrows are dug, the birds begin visiting each other in the many burrows. As the razor-bills and guillenots arrive, they share the burrows, usually remaining closer to the entrance than the early-arriving puffins. Sometimes fulmars, gull-like seabirds, lay their eggs at the entrance to the burrows. But the puffins are hospitable and step carefully around the fulmar's eggs.

For 38 days after egg laying, the puffins share incubation duties, sitting while the other feeds and then switching roles. Sometimes both birds leave the nest, leaving the egg untended for hours.

When the young puffin hatches, both of the parents share the feeding duties. They fly to the water and fish until they have as many as 20 small fish, which they hold in their fat bills; then they return to the nest to feed the little one. The meal is cold, raw fish—nothing warmed and predigested as so many other seabirds give their young.

Seven weeks pass before the fledgling puffin is ready to emerge. By then the parents are seldom seen. The fledgling, which could walk the day it hatched, must learn to find its way about outside the burrow, then learn to swim and fly entirely on its own. By midsummer the young puffin launches itself into the air, flying out to sea to return the following spring.

*The **Atlantic puffin** is not related to the penguins of the Southern Hemisphere despite its black-and-white feathers.*

Bill to bill, these pelicans struggle over a morsel of food. If approached while nesting they defensively clap their bills emitting a peculiar sound.

Before Our Eyes . . .

. . . the brown pelican is on its way to becoming an endangered species. Even as we talk about saving wildlife from extinction at the hands of man, we are doing to the brown pelican what has been done to countless other species: wiping them off the face of the earth. This time we are using chemical warfare.

But the pelicans' tolerance for pesticides, like that of many other animals, is not particularly high. The worst offender against the pelican that man has offered is DDT and other chlorinated hydrocarbons. These pesticides are sprayed on plants, but most of the deadly poison finds its way into the rivers, which flow to the sea. Plankton is therefore contaminated. The plankton is eaten by larger organisms, and each of these steps along the food chain carries the sum total of the pesticide in all previous steps. By the time fish eat the pesticide-infested organism that is its food, the concentration may be as high as 20 or 30 parts per million. The U.S. Food and Drug Administration says marine fish with anything above seven parts of DDT per million is not fit for human consumption.

The California subspecies of the brown pelican was a common breeding bird along that state's coast as recently as the early 1960s. Today breeding brown pelicans are less common there. Biologists saw it coming, and they even saw why it was coming. In 1966, for example, they found birds with 84 parts of DDT in their tissues for every million parts of tissue. Two years later there were no more breeding pelicans in California. What had happened, scientists learned, was that the concentrations of DDT affected the pelican's ability to produce eggshells strong enough to protect the embryo. Their eggs were breaking even as they laid them because the shells were so thin. In 1969 researchers went into the field and counted 298 nests in one colony of pelicans at Anacapa, an island off California. Unbroken eggs were found in only 12. Another 50 or so contained a single shattered egg. Another colony of 339 nests yielded 19 unbroken eggs. In both colonies, most nests with viable eggs held only one. Healthy pelicans usually lay three or four eggs at a time. The cause in each case was traceable to DDT.

Pesticides used in the past 35 years since DDT came on the scene have been building up, and much will be ingested somehow by various living things. So as the brown pelicans of California die off, they will have fewer replacements growing up behind them, and there may soon be no West Coast population of these birds. Already they are gone from Louisiana, where the brown pelican is the state bird.

Soon the brown pelican may make the list—the official endangered species list.

The **brown pelican** *(right) has dwindled in population because of DDT poisoning. Their ungainly chicks (top) will grow to have a six-and-a-half foot wingspread. The pelican's nest (bottom) is made up of sticks and dried grasses.*

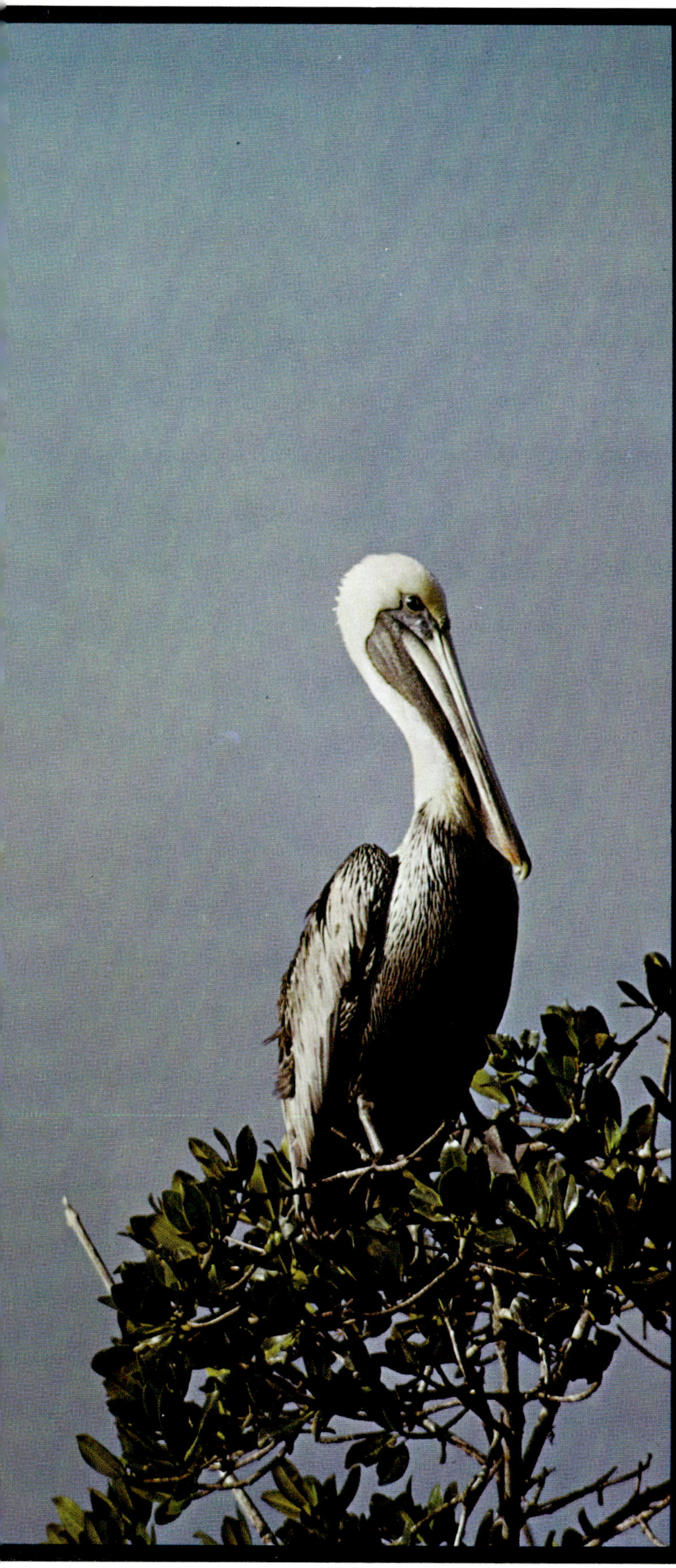

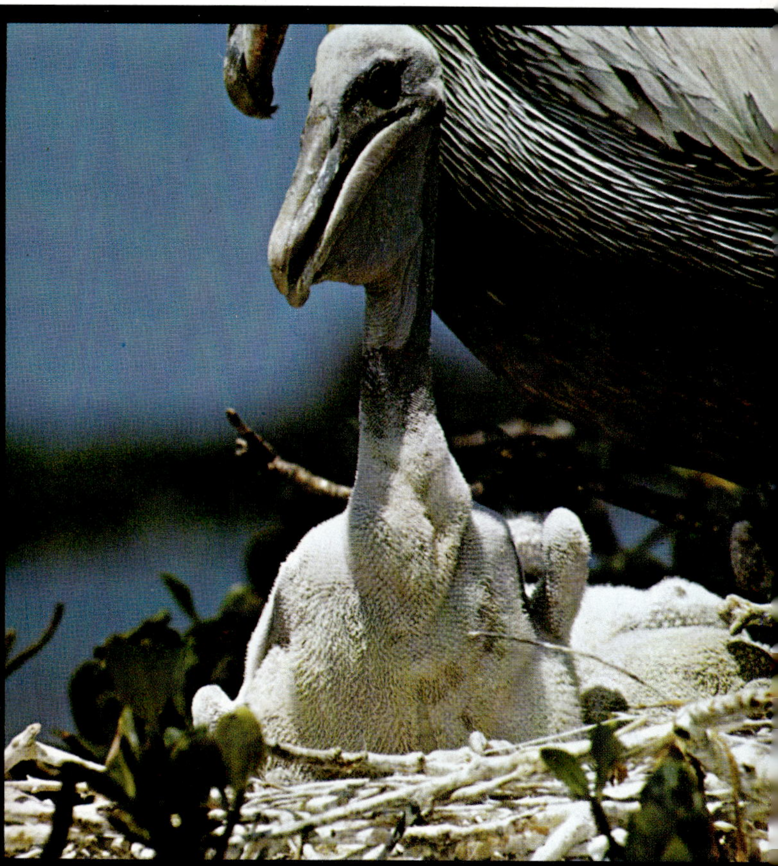

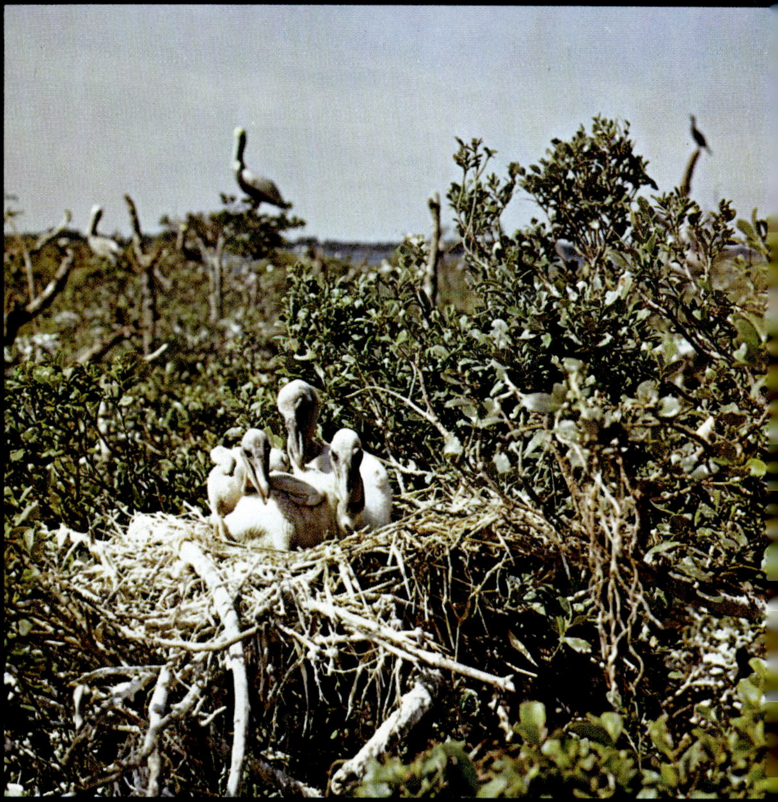

The Wanderer

The albatross is a good omen to seamen and to kill one is to bring bad luck. When the Ancient Mariner of Samuel Taylor Coleridge's Rime shot one of these winged wanderers, his shipmates hung it around his neck as a reminder that it was his fault that they had become becalmed in mid-ocean.

In 1936 the ship of the French polar explorer Jean Charcot, the *Pourquoi Pas*, ran aground in a storm and sank. When Charcot was alone on board, he went down to his cabin, picked up a wounded albatross that he had been nursing for weeks, and released the animal which had recovered and was able to fly in the strong wind. Then Charcot returned to his cabin and disappeared with his vessel.

The albatross is the largest of the seabirds, wandering over vast tracts of oceans, out of sight of land for months or even years at a time. They travel tens of thousands of miles wherever the winds are strong and constant, riding on wings extending as much as 12 feet across and catching updrafts that help keep them aloft for hours of gliding.

With a heavy body, long, slender wings, and legs set far back, the albatross is such a clumsy creature when it does alight, that it has won the name "gooney bird." The gooneys come to land only for breeding purposes and subsequently to raise their young. Coming in to land, the albatross is so awkward it may fall over, collide with something, or even accidentally somersault.

Mating ritual. *With head and neck stretched skyward these albatrosses approach breast-to breast and perform a ritualistic mating dance, circling each other and bobbing to and fro.*

Incubation. *After a period of two and one half months of alternately incubating the eggs, the parent albatrosses will be rewarded with two hungry chicks, which will be fed regurgitated food.*

One by one these great fliers come to land, until whole colonies have gathered to find a mate, breed, and raise the young. It takes close to a year from egg laying to takeoff of the young. The egg, a monster weighing in at three-quarters of a pound, is laid in a nest built or rebuilt largely by the male. Nesting follows the highly ritualistic dance of the gooneys in which the birds come together, breast-to-breast, bills pointed skyward, and move about with an awkward sort of grace. Both birds take turns incubating the egg for two and a half months. And they share in feeding regurgitated food to the young bird, which weighs in at 12 ounces at hatching and may weigh more than six pounds at the end of its first month. The hatchling is covered with silky white down, followed a month later by dense gray brown down. Feathers come later. The albatross only begins to mate at the age of six years and produces one egg every other year. The slow rate of renewal makes the species very vulnerable.

The young birds as well as the eggs are considered good eating by man and skua, a pelagic scavenger bird. And when the young birds reach the water, they are fair game for sharks. One shark caught off Midway Island was found with seven gooneys in its stomach; knowing sharks, this may have been a meal of sick or dead birds. The gooney's principal defense is a foul-smelling, sticky oil they expel from their stomachs.

The albatross is certainly one of the finest fliers in the bird world, yet it takes many long hours of practice for the young gooney to get off the ground and remain airborne even for a few seconds. It tries over and over, launching itself from heights or running to catch the wind. If there is no wind, even the adult albatross caught on the ground can not get airborne. When the juvenile bird at last takes off and flies out to sea, it is about a year old. At this point the ungainly gooney becomes a graceful albatross. It remains a juvenile for years, reaching sexual maturity at about six years of age if it is Laysan albatross or close to nine years of age if it is a wandering albatross. With good luck an albatross may live to be 30.

*The **baby blackfoot albatross** eats predigested food. The young bird reaches deep into the throat of its parent for freshly caught and eaten fish, which is already partly digested and warm.*

Sperm Whales

The sperm whale is the least graceful mammal in the sea. It is without a dorsal fin, always shows its tail when diving, blows its "spout" at a 45° angle to the left and forward, and has a monsterously ugly cylinder-shaped head. Though its skull is only a very small part of that head, the sperm whale has the largest brain of any living creature. When it is fooling around, its speed is three to four knots; its cruising speed is eight knots; it occasionally swims at twelve knots, and when jumping clear of the water, it reaches 22 knots.

The sperm whale has nothing to fear from the attacks of the few animals that would dare to prey upon it. Its powerful tail would kill or seriously harm the sharks or orcas that would try to approach it. Herds of sperm whales have been seen gathering in a circle around a wounded calf with their heads facing toward the center and their broad and powerful flukes slapping the water to chase off a pod of orcas.

When disturbed by man, a herd of sperm whales is slow to realize the danger, and it is during that period of surprise that whalers take their catch. But when the whales understand their danger, they display an extraordinary show of intelligence, solidarity, communication, and defensive strategy.

Another example of cooperative behavior among these great toothed whales is when a pod is at rest, lolling about in calm seas. Several members of the pod remain alert, stationed in various places, and patrol around the edges of the group, ready to warn of approaching danger.

Sperm whales are polygamous; the great males, 50 to 60 feet in length, form harems of many cows which are always slightly smaller, ranging from 35 to 45 feet. Young of both sexes often travel with these harems. Smaller pods of bachelor bulls run together, and older bulls may remain by themselves. The sperms mate during the first half of the year in temperate and warm water. Calves 13 feet long, weighing one ton, are born after 14 months and nurse for two years. In that time, a calf may double its length to about 25 feet. Its mother and other females of the pod watch over the calf, especially if it is sick or injured. Sperm whales become sexually mature at about nine years of age and continue to grow for up to 25 years, at which time the bulls are ready to become sultans of a harem. Some sperm whales have been found to reach the age of 70.

Sperm whales rove the oceans of the world, the harems generally remaining within the temperate and warmer waters in search of the squid and cuttlefish that make up their diet. Only males, either old solitary ones or groups of bachelors, migrate to polar waters. They are known to dive eventually to depths of a mile and more and frequently remain underwater for 50 minutes. When they return to the surface, they may "pant" briefly by breathing several times a minute for a few minutes, then dive again.

These great whales are largest of the toothed cetaceans. Their large blunt heads are full of a waxlike substance called spermaceti. The spermaceti is most probably part of a transducer system allowing echolocation in low frequency and at great distances in a plane perpendicular to the axis of the head.

Spermaceti has been one of the reasons the sperm whale has been a victim of whalers. The waxlike substance, which remains liquid at normal body temperatures and becomes a solid when cooled, was very highly regarded as a lubricant and fuel. The foot-thick layer of blubber that sperms carry is another source of fine whale oil, but this oil, very different from that obtained from other whales, cannot be transformed into edible fat by hydrogenation. Their bones and teeth have been widely used for a variety of crafts.

Characteristic blow. Sperm whales pass (opposite, top), some blowing their characteristic blow, forward and off to one side.

Sperm whales (opposite, bottom), largest of the toothed whales, grow to 60 feet. Dwarfing the diver (below) their head and jaws are unique among cetaceans. Only the lower jaw of the sperm whale possesses visible teeth. The upper jaw contains sockets into which fit the teeth from below.

The Walruses

When translated, the scientific name of the walrus means "he who walks with his teeth." The tusks of the walrus are canine teeth that may grow up to three-and-a-half feet long. The animal has been seen using its tusks to climb on an icefloe by digging them into the ice at head level, then pulling itself up on them. But principally the walrus uses its tusks for fighting and for feeding.

To dig clams and mussels, walruses may dive as deep as 300 feet to the gravelly sea floor, but it is highly debatable if they rake them up with their tusks. Trying to systematically justify anatomical anomalies like the tusks of the walrus or the spear of the narwhal is just as meaningless as explaining the reasons for Cyrano's nose. Walruses are able to remain underwater 15 minutes or more at a time, but they stay exactly four minutes at each dive when they are migrating. They cruise at five knots and can reach a speed of 15 miles an hour. Their seeming clumsiness on land melts into a graceful flowing motion in the water. But as soon as they climb onto ice floes from the sea, every movement seems to take a great effort, adding to the apparent awkwardness of these big one-and-a-half-ton bags of meat.

Walruses are born out of water after a gestation period of almost a year. They quickly learn to swim and dive with their mothers as teachers. But as infants, they have no tusks. They remain sucklings for up to two years, but supplement their mother's milk by whatever food they can find and chew on the bottom. Mother and calf remain very close during this time; the mother not only teaches the calf, but also protects it from its principal predators, the polar bear, the orca, and man. When they are old enough, with tusks perhaps three or four inches long, the young walruses are on their own. They reach sexual maturity at five to six years of age.

It was believed that walruses lived exclusively upon clams and shellfish. In fact, the stomachs of walruses very rarely contain shellfish and much more often ascidians, starfish, urchins and other bottom fauna from polar waters. They add to their diet clams that are loosely attached on muddy bottoms and that they pick up without the help of tusks. They crush them with their powerful teeth. Their coarse and bristly whiskers are used to "feel" their food in waters that are generally very turbid.

Walruses generally live in herds numbering into the hundreds, resting on huge ice floes and frequently crowding in on them so much they rest their tusks and their bodies on each other. As they lie around at rest, they frequently rub themselves to get rid of skin parasites. They bellow, growl, roar; underwater they clap their teeth in serials, and sometimes makes a peculiar bell-like sound. In a family apart from all the other pinnipeds, walruses have thick skin and sparse and patchy fur. The inch-thick skin, covering six inches or more of blubber, makes them fair game for Eskimos and northern Indians; the hulls of their umiac boats are made of female skins (the male's skin cannot be used because it has many tusk holes resulting from fights).

On land walruses can turn their hindlimbs forward the way sea lions and fur seals do, enabling them to move quickly for such ponderous beasts, up to running speed for a man. During their regular migrations north to the high arctic in summer, they often ride ice floes part of the way.

Two distinct groups of walruses are recognized. They are considered the same species, but separate subspecies. The Atlantic walrus, found in the northernmost reaches of the North Atlantic and north of the Soviet Union in the Arctic Ocean, has a narrower skull and shorter tusks than the Pacific race, which lives in the Bering Sea and northward.

Walruses are social animals (above). They are noted for their long canine teeth (below), used for feeding, fighting and to pull themselves onto icefloes.

Harbor Seals

There used to be a $5 bounty on harbor seals in some areas of the east coast of the United States. To collect the bounty, a person had to bring in the nose, cut from the seal's carcass. The bounty was first put into effect many years ago at the urging of commercial fishermen who believed the seals were depredating marketable fish as well as damaging their nets and pots. While it is true that harbor seals eat some fish, they also eat molluscs, squid, octopods, and crustaceans, most of them of no commercial value. Eventually the various states realized the bounty was responsible for a needless slaughter and the bounty law was repealed a few years ago. Nevertheless the seals have continued to disappear from many areas where they once were common, draped over rocks or cavorting in the shallows near shore.

A major cause of the disappearance has been man's continued encroachment on the seal's habitat. The great increase of pleasure craft in coastal waters has apparently driven the seals to quieter waters.

Harbor seals give birth to their pups in late spring and early summer, depending on the latitude and local conditions. The pups are born one to a mother, close to the high tide mark. They have a coat of long, white hair they shed either just before or just after being born. Beneath the long coat is a shorter, tougher coat of white that darkens slowly with age until the mature seal is dark gray with white spots. The pups can swim almost as soon as they are born. They suckle underwater for about three weeks, and by then they are strong enough to swim and seek out their food especially crustaceans. By the following spring, the yearling pups form groups of their own, while older seals of both sexes gather in separate herds. Often a bigger, older male stands watch over the several herds from a distance. The seals reach sexual maturity at about three years of age. They mate immediately after their spring molt. The pups at birth are about three feet long and weigh close to 35 pounds. Full-grown males may weigh up to 350 pounds and be more than six feet long, while females are slightly smaller and lighter.

Harbor seals are members of the earless, or true, seal family, and are generally more aquatic than the eared seals except in having to come ashore to sleep. Sea lions and fur seals often sleep in the water. Harbor seals, however, mate in the water, and are adept swimmers and divers, using their hindlimbs sideways for swimming at speeds up to 15 knots. They can remain underwater for up to 45 minutes at a time. On land, they are unable to use their hindlimbs because they do not hinge forward; they move pretty swiftly, though, arching their backs rhythmically, as caterpillars do.

Harbor seals have been hunted for thousands of years. In more recent years, Eskimos and northern Indians of Alaska have shot the seals which they feel are depredating the salmon fishery. In Germany and Iceland the white-coated seal pups are sought to be made into fur coats. And in Japan the seals are exploited by commercial interests which convert their skins into hides, blubber into oil, flesh into pet food, and bones and internal organs into fertilizer.

One subspecies of harbor seal has adapted to a freshwater existence in northern Quebec. Others, numbering perhaps 400,000 in four subspecies, are found around the world in northern Europe, Asia, and North America.

Harbor seals range from temperate to polar waters (opposite page). A harbor seal pup (below) can swim soon after birth and dive for 45 minutes.

Blue Whales

Great size is seldom an indication of ferocity. Quite the contrary, it seems as if the larger the animal, the more peaceful it is. Nearly 200 million years ago there was a dinosaur called *Diplodocus* that grew to a length of nearly 100 feet and may have weighed 40 tons. It lived in marshes where its great weight could be supported by the waters. *Diplodocus* was vegetarian and munched plant life in the shallows.

Today there is in existence an animal even longer and much heavier than *Diplodocus*. The blue whale reaches a maximum length of close to 118 feet and weighs in at 150 tons. At birth, a blue whale is about 23 feet long and weighs close to seven tons. Its mother's milk helps it grow at the phenomenal rate of one and a half inches and 200 pounds every day for the seven or eight months it nurses. During that time, it daily consumes close to 300 pounds of fat-rich milk with a consistency of runny cheese which has no fishy flavor. Whales' milk is the richest of all mammals. The diet puts on the scrawny calf a generous protective coat of blubber. By the time it is weaned, the calf may weigh 23 tons and measure up to 53 feet in length.

Once weaned, blue whales feed mostly on plankton and small fish, and during the two or three months they spend in the antarctic annually, or krill, a one-to-two-inch shrimp-like crustacean which is plentiful in austral polar waters. They filter the krill on their more than 300 baleen plates. The stomach of a blue whale has been found containing one ton of krill, and it is estimated that they are able to consume about five tons of krill per day.

Today there are so few blue whales left it is difficult to learn more about them. Relatively little is known that can help us save these giants from extinction. We know they arrive singly, in pairs, or in small family groups of three on the feeding grounds with layers of blubber thinned out to only 15 percent of their total weight. By autumn, after feeding on the krill with its high fat content, the whales have fattened up until their blubber equals 30 percent of their total weight. During their pelagic migrations, the blues cruise at speeds of up to 15 knots with bursts up to 20 knots for 10 minutes at a time. They can dive to 1000 feet or more but usually remain in shallower water where krill gathers in vast swarms. Krill and other plankton sometimes go deeper, however, and the blues go where the food is.

In autumn, when the whales leave their feeding grounds, they are fattened for their long partial fast while they travel and sojourn in warmer waters. In these warmer waters, a cow gives birth to a single calf no more often than once every three years after a gestation period of 10 to 11 months. No one seems to know for sure when blue whales reach sexual maturity. Some say at eight years, others say at 25 years. They are known to live to 90 years. Age is determined in many species of whales by counting the layers of ear wax that are deposited each year, a method reminiscent of counting tree rings.

Some persons have claimed the northern race of blue whales is not the same as the antarctic species and thus not subject to the protection they need. The protection given by the International Whaling Commission prohibits taking any blue whales. The job of enforcing the regulations, even with observers aboard whaling ships, is extremely difficult and there is little doubt that some blues are still being killed each year. Whalers may claim difficulty in identifying the species of whale they have shot because of fog or heavy seas, but blues are usually easily recognized by their pear-shaped blow that billows up some 30 to 50 feet into the air. Their size, their configuration, and their coloration are all additional clues to identification.

Generally, blue whales are blue with blotchy lighter markings along their sides. On the underside, the pale green blue is sometimes heavily colored with diatoms, giving them a yellow hue beneath and yet another name, sulphur-bottom.

Some optimistic observers feel that full protection would enable the blues to increase their numbers to the maximum yield level in the year 2100 which would forestall extinction of their species. But the numbers today are so few in such a vast volume of water that they may have difficulty finding each other as they travel singly or in pairs rather than in large pods. There may be as few as 1000 in the northern hemisphere and only two or three thousand more in the much broader reaches of the southern seas.

Chapter VI. Man, The Despoiler

In Scandanavian archeological digs, the remains of whale bones and stone harpoon heads dating back to about 2600 B.C. have been found. The fishermen of Crete hunted dolphins hundreds of years earlier. The Phoenicians had a whale industry by 1000 B.C., and the Basques had a full scale whale fishery in the Bay of Biscay by A.D. 1200.

In all the long history of whaling, however, the first scientific study of the life and death of any whale species was not made until 1929. Some excellent but incomplete studies were made in the nineteenth century. The long delay was partly a result of the diffi-

> **"It is only since man's conquest of fossil energy that things have changed and that for man to survive, he has to protect nature instead of fighting it."**

culty of studying these sea creatures and partly the result of man's traditional belief that to survive, he had to fight nature: he was weak and nature was hostile. It is only since man's conquest of fossil energy that things have changed and that to survive man has to protect nature instead of fighting it. This has affected the lives of hundreds of species of mammals and birds in the sea. Many have been decimated, some have been extirpated, and if they are known at all today, it is only through photographs and old engravings. Man ranks as the most thorough, all-embracing predator of all time.

In the scheme of things there are predators and there are prey. These roles are often reversed so that an animal that is a predator at one time becomes prey under other circumstances. Man raises sheep, cows, and chickens for food. We also hunt wild deer, rabbits, and birds to kill them for food and so-called sport. The quantity killed by hunting decreases rapidly although the number of hunters increases because there are fewer and fewer animals to kill. On the other hand, we are only subject to predatory bacteria, viruses, fungi, and roundworms. There was a time when the cave man was an easy prey, but today he is the ultimate predator.

Man despoils the planet and especially the rivers and the oceans in other ways. The human population is growing exponentially and efforts to control it still have to demonstrate any efficiency. Man, like birds and mammals, needs water and competes successfully with wild animals for the occupation of river banks or seashores. In choosing the sites for his exploding cities, man eliminates the habitats of billions of living things. A filled-in paved-over marsh, sprouting ranch houses, is hardly a place for herons and ducks to nest.

Whether by shotgun, harpoon, or cannon, whether by poison, overcrowding, or habitat destruction, there are no more Labrador ducks or great auks left in the world. The Eskimo curlew is all but gone. The blue whale and other rorquals may be on their way out. Steller's sea cow, seen in eighteenth-century Alaska, never made it to the nineteenth century. Eight hundred living species have been eradicated by man's criminal carelessness between 1900 and 1960.

*The **plight of the whale** is dramatically illustrated in the scene at the right. For centuries whales have been hunted and exploited by man and now a few species are in danger of becoming extinct. Only international agreements between whaling nations have saved some of these mammoth creatures from total annihilation.*

101

Eskimo Hunters

Peaceful and shy when it is left alone, the most dangerous animal in the arctic to man is a wounded walrus, according to the Eskimos who hunt them. With its tusks and its massive size and weight, along with its agility in the water, the walrus can hook its tusks over the edge of a umiak or kayak and dump his tormentor into the frigid waters. The hunter then becomes the hunted. A typical attack begins with the wounding of the animal, who enraged and in pain often turns on the flimsy kayak or umiak. If the occupants are not quick enough, the walrus is on them, upsetting their boat and dumping them into the water where the walrus can deal with them more easily. If sufficiently provoked he may continue by picking a victim, then plunging its tusks through the man's chest or back. Then the walrus goes back to the surface to pick another victim and smother him under a ton and a half of flesh and bone, before piercing him with tusks.

When we asked for the time the last such accident occurred, no one in the villages could remember. Eskimos continue to hunt the walrus today. Under licenses granted by the Russian, the Canadian and U.S. governments, they take a quota of walruses each year. Thanks to these regulations their numbers that had dropped dramatically and dangerously are now increasing very slowly. The Eskimos kill only what they need and utilize virtually every bit of the animals they take. The dark red, myoglobin-rich flesh is food for them and their dogs. The oil they wrest from the blubber is fuel for their stoves and lamps. Skins are used for the hulls of their boats and for their ropes. Windows that admit light into their huts and igloos are made from the translucent intestines. And the tusks are used to line the keels of their boats and are fashioned into knives, needles, beads, buckles, sun goggles, snow cutters, and a host of other items that help them survive.

The hunt is carried out by men in umiaks, with a crew of four to six men, and now usually powered by outboard motors. As soon as a herd of walruses is seen, the motor is turned down to slow speed and later stopped, and paddles are used until the umiak gets within shooting distance. Harpoons are also ready to penetrate the skin and blubber and keep the walrus from sinking out of sight or escaping under the ice, if it is only wounded at first. Some Eskimos use buoys made of skin which they attach to the harpoons, to keep the walrus from submerging for long and to tire the animal quickly.

Walruses (left) are hunted by eskimoes from umiaks made of walrus hide (below). The skin (above) is being prepared for the construction of a umiak.

Walruses existed by the millions a few centuries ago. At the turn of the century they were less than 60,000. Today there are hardly 100,000, if we count the populations of the Atlantic, Pacific, and Arctic oceans. This is still a highly critical number. The extinction of the walrus would mean the end of the Eskimo civilization.

*As one of the more vocal of all the whales, the **beluga**, is audible even while submerged. The word beluga means white in Russian.*

White Whales of the North

The Indians of the Hudson Bay region in northern Canada believe that the flukes of an unborn beluga porpoise emerge from the mother anytime from four to six weeks before birth so that the fetus can practise swimming. That way, say the Indians, the "white whale" is ready for life in the sea as soon as it is born. This naïve legend only proves that the Indians know that belugas are good swimmers at birth.

Belguas are the only cetaceans that are normally white and their name is the Russian word for "white." They live in the northernmost reaches of the world, migrating southward in the Western Hemisphere only down to the Gulf of St. Lawrence. They grow to a maximum length of about 16 feet and dwell in the coastal waters of northern Canada, Alaska, Siberia, Russia, and Norway, at times running many miles up rivers. With a top speed of a little more than 10 knots, they are fair game for the swifter, voracious orca which attack and kill but, Indians say, never eat any of the cows that fall victim to them.

The other predator on belugas is man. The Hudson Bay Company sent hunters out for the white whales in 1688, and today man continues to hunt the belugas with guns, harpoons, and nets. Russian whalers capture belguas in beach seines as the animals travel in small family groups of five to ten, and sometimes they capture individuals from larger groups of 50 or 100 that are divided according to sex.

When the Indians spot a three-foot blow of a beluga, they know that there are probably a number of whales in the vicinity. They launch their boats to give chase in the ice-choked waters, and once they have run the belugas down, they harpoon them and tow ashore all they have been able to kill. Hunting belugas has been a way of life for these Indians for many generations, so they know how to utilize the animal to best advantage. The skin, for example, is quite thick and is tanned into a fine leather. The flesh is eaten by them or by their dogs, and they render oil from the blubber and other parts. Commercial hunters in many parts of the arctic world have almost wiped out the belugas.

The Canadian government regularly surveys the populations of belugas in its waters by helicopter observations. The light-colored porpoises show clearly in the dark waters when viewed from the air. It is estimated there are only about 5000 beluga whales left in the world, a very critical number.

Eskimo fisherman is carving up a beluga whale. Making full use of the by-products, the skin will be used for leather, the flesh eaten, and oil rendered from the blubber. The populations of belugas are becoming critically low as a result of fishing pressure.

"Thar She Blows"

In the nineteenth century, if a man or boy could put up with maggot-infested food, harsh discipline, and three years away from home, he could go to sea aboard a whaling ship and earn about 20 cents a day. In spite of all these discouraging aspects, in 1846 there were more than 700 American ships hunting sperm whales. (The American fleet then was almost double that of all other countries combined.) New Bedford, Massachusetts, was the whaling capital of the world. More than 100,000 barrels of sperm oil a year was produced by these crews. The ship owners and the captains they hired got rich even if the crews did not. Crews usually signed on for a share of the earnings—often only 1/200 of the take—and they could lose even that pittance as punishment for minor infractions of ship's rules. Bad as these conditions were, the whales fared worse.

The man who first spotted a whale might get a sack of tobacco as a bonus. Spying the spout of a whale he would call, "Thar she blows." And the reply would come from the skipper, "Where away? What d'ye call her?" "Sperm whale. Starb'rd bow, two miles or more to wind'ard," the lookout would call back. And the chase was on. The ship would draw as close as possible to the whale, then the 30-foot, narrow-beamed shallops would be quickly lowered. The officer would steer the boat from the stern, and the harpooner, standing in the bow, would be ready to plunge his iron into the whale as soon as the boat was in position. With a shaft of iron sticking into it, the whale would take off, just under the surface or diving deep, and up to 4000 feet of inch-and-a-half manila line would whir out of the boat, charring the wood as it spun out. The mate would leave his steering oar astern and exchange places with the harpooner, to plunge his more delicate, scalpel-sharp lance into the whale's heart. Sometimes a skillful mate could make a quick clean kill, but more often the death throes of the whale lasted many minutes or even hours. Men were maimed or killed in some of these encounters; boats were splintered by the whale's flukes. Once the whale was dead, the ship was brought alongside and cutting in began while the animal lay in the water. First the jaw, then the head was cut and left afloat for the end. Blubber was stripped off with flensing knives that had long handles, and the long strips of peeled fat were then chopped into smaller chunks to be boiled in the try-pots on deck for the oil. The head, if not too heavy, was painstakingly hoisted on deck. The reservoir of waxy spermaceti in the bulbous head of the sperm whale was tapped and the product stowed separately from the other oils tryed from the carcass because of its waxy rather than oily character. Before the carcass was abandoned, spears and lances were driven into the whale's belly and carefully smelled to learn if ambergris was present.

When a crewman had a bit of spare time, he might take one of the sperm whale's teeth and carve pictures on it.

If a ship took a rorqual instead of a sperm whale, the baleen could be used to make corset stays, which were needed for the fashions of the day.

*In the fury of battle between mariners and whale, a 60-foot **sperm whale** hurtles out of the water spelling doom for any hapless whalers below.*

Whaling Today

They don't call, "Thar she blows!" anymore. Instead it's a nod or a gesture with an arm.

Whaling is a whole different business now, and there are few remnants of the picturesque saga of 100 years and more ago.

Helicopter teams spot the pods of whales and radio information to a mother ship. This is an elaborately equipped floating factory designed specifically for rendering and processing everything that can be obtained from a whale. This vessel is like a mother hen, and her brood of "chickens" are catcher and "supercatcher" boats, up to 90 feet long, and operate under the direction of a gunner. His weapon is a swivel gun loaded with a 150- to 200-pound harpoon carrying an explosive head, timed to detonate three to five seconds after penetrating the whale's body. A line is attached to the harpoon at one end and to a power winch at the other. In case the first explosive harpoon does not kill, the gunner is ready with a second one without a line on it. This usually finishes the whale. Before the great beast can sink, the crewmen on board the catcher whip a length of cable around the whale's tail, punch a hole into the animal's abdomen, and pump sev-

Demise of a whale. Sophisticated armory impails the victim (above and left). A sperm whale yields her unborn calf to the whalers (lower left and right). Nothing is wasted, not even the young. Estimates are that one half of the female whales killed are pregnant.

eral thousand cubic feet of compressed air into the carcass to keep it afloat. A buoy equipped with a flag and radio transmitter identifies the dead whale's owner and calls it to the attention of the factory ship, which may be as much as 100 miles away.

As soon as the dead whale is alongside the factory ship, a grab hauls the whale aboard through its open stern. The flensers, using flencing knives, go to work stripping the inches-thick blubber from the animal. The 10-by-1½-foot strips of blubber, as in the old days, are cut into smaller pieces, then machines feed the chunks below decks into 600 psi pressure cookers called try-pots where the blubber renders its oil. Meanwhile the carcass is flensed; meat is sent to the canning department, the bones are fragmented with electric saws, also to be boiled in try-pots to yield up their oil. A hundred years ago a whaling ship could count on capturing 35 to 40 whales in a three-year trip, netting them 800 barrels of oil. Twenty years ago, at the peak of industrial whaling, as many whales were processed in two weeks, and each eight-month trip netted a million barrels of oil.

Today Russian and Japanese whalers are responsible for nearly 90 percent of the world's whale catch. The principal remaining whaling is carried out by Norway and South Africa. At the current rate, they will shortly be out of business. But they will leave behind a legacy of wasted resource for there will be so few whales that it is doubtful they could reestablish a viable population.

A moratorium on all whale killing would be the only chance the whales may be saved. Scientists have called for a ten-year respite for the whales as a minimum requirement. But it would take a lot more time, perhaps a hundred years, for such varieties as the blue whale, which is almost extinct and has a slow reproductive rate, to recover.

After hauling a baleen whale out (above) sharp knives and winches will dissect the mammoth to a naked carcass in short time (below).

Harp Seal Horror Story

"Hit 'em in the nose."
"No, no, hit 'em on the head."
"I say kicking's best."
"Shooting's surer."

There are six ways to kill a baby harp seal for its pelt. Gaffing and clubbing seem to be among the most popular, possibly because the equipment needed is easiest and cheapest to get. In gaffing, the seal hunter stalks the baby seal, which moves slowly or not at all. Then, with a chunk of wood that has a heavy hook and a spike in it, the hunter raps the pup on the head with the hook and drives the spike into its brain. In clubbing, they use wooden clubs like baseball bats. It may take four or five or more blows to render the pup unconscious or dead.

There are other ways like kicking the pup in the face, then quickly rolling it on its back before it can recover and slitting its throat. But this works only with pups; adult seals will try to bite the leg of their assailants. Shooting is used principally for adults trying to get away. Unfortunately for every adult harp seal killed and recovered, five others, wounded, get away into the water. Drowning is an effective method of capture, but it takes longer—half an hour or more sometimes. Hunters set nets or special traps for the seals and leave them underwater where they are caught and held until they are dead. And then there is a long-lining, which is very much like fishing. Baited hooks are lowered in the waters where hungry harp seals may take the fish bait and drown on their own blood after swallowing the sharp hook.

Using one or several of these methods, a good seal hunter can kill 125 or more seals a day. He works for a little more than a month of the year in the Gulf of St. Lawrence. If he is one of the big commercial hunters, he can make over $1000 just for killing seals. But many of the hunters take fewer seals and make about $500 a season. What kind of men are these hunters? They are, very often, simple men, fighting against poverty and enduring bitter cold in a bloody business. Many have said, "If there was something else I could do at this time of year, I wouldn't kill seals."

Harp seals are found around the polar seas of the north—off northern Norway, in the

Nosing around. *A harp seal (above) pokes its head into a hole in the ice.*

White Sea, and in the Canadian arctic and subarctic. They spend much of their year at sea. In winter the harp seals between Baffin Island and Greenland and from Hudson Bay move south to the Gulf of St. Lawrence to give birth to their pups on the ice floes. The pups at birth are about 30 inches long and weigh about 25 pounds. They have yellowish curly fur which changes to white in a few days. Feeding on mother's milk, they fatten quickly, building layers of blubber for protection against the cold. At one month of age, their coats are half an inch thick and pale gray. By their first birthday, the pups have darkened and spots have begun to appear. It is during that first year that the hunters prefer to get the pups. And because harp seals have no enemies, other than man, they are usually easily approached. Often the pups "freeze," lying perfectly still with eyes closed and heads drawn in, presumably to blend with the icy background. A hunter can pick a pup up with no trouble. Mother seals make an attempt to protect their young, but they want to save their own skins too. So they frequently make a threat and disappear into the leads of water between ice floes, when their threats are ignored.

The world population of harp seals is about five million. In Canada surveys seem to indicate the seals herds in that country have been reduced by half in recent years. Hunting off Norway and in the White Sea also continues unabated.

Alerted seals. *Fear of man causes these harp seals to withdraw from the crawling photographer.*

Chapter VII. Man, The Restorer

In 1905 conservationist Theodore Roosevelt was president of the United States.

In 1905, Audubon Warden Guy Bradley sat on the porch of his house in Flamingo, the southernmost part of peninsular Florida, looking out over the waters of Florida Bay. Around him were birds of many species that were his to protect for the people of the en-

> **"The International Whaling Commission has not set sufficiently realistic standards."**

tire country. He saw a schooner belonging to Walter Smith, known as a hunter of the plumes, or aigrettes, of the protected snowy and American egrets, anchored off Oyster Key. As he got into his rowboat, he heard shots from the Everglades nearby. When he pulled alongside the schooner, Smith's son and a friend of the Smiths climbed aboard, each with two dead egrets.

"What do you want?" Smith called to Bradley. "I want to arrest your son," he replied. "Well, if you want him you have to come aboard this boat and take him," said Smith. As Bradley manuveured to tie up to the schooner, he told Smith, "I will come aboard." Smith fired a bullet from his Winchester rifle into Bradley's chest killing him instantly. The warden's body was found 24 hours later, adrift in Florida Bay. His murder caused an uproar, but the grand jury refused to indict the plume-hunter. His neighbors did, though, when they burned his house down. Even as Bradley's killing built up impetus for the conservation movement, Bradley's replacement, Columbus C. MacLeod, was shot, probably by plume poachers.

Since then, many other conservation organizations have been established to conserve the remaining species of many plants and animals. One such organization carrying the official blessing of national governments but little else, is the International Whaling Commission, founded in the mid-1950s to regulate whale hunting. Recently this Commission rejected a proposal for a ten-year moratorium, although it obtained eleven pro votes and only five opposed with one abstention, under the shameful pretext that such a decision needed a 75 percent majority!

Most of the whaling nations are members of the commission. While the commission's job was to ascertain the maximum allowable whale catch without decimating the various species, it has failed to set sufficiently realistic standards and has been unable to enforce the insufficient ones it has set. Meeting once a year, the commission eventually got all its signatories to stop killing blue whales. Two nonmembers, Peru and Chile, continued to kill the rare whales until recently. The Blue Whale Unit, an artificial measure, was designated by the commission to establish how many whales could be taken by each nation. Two finbacks equal one BWU, but six sei whales equal one BWU, putting enormous pressure on the smaller sei whale and drastically affecting its worldwide population.

Pressure from the commercial whaling interests on the commission and the inability of the commission to enforce its decisions have meant, in general, that the limits they have set each year have been far in excess of the limits proposed by scientists, and that would have effectively avoided the extinction of most whales in a few years.

Rescue operation. *One of hundreds of volunteers, who turned out when an offshore oil well in California blew out, carries a badly oiled but still living western grebe to a cleaning station.*

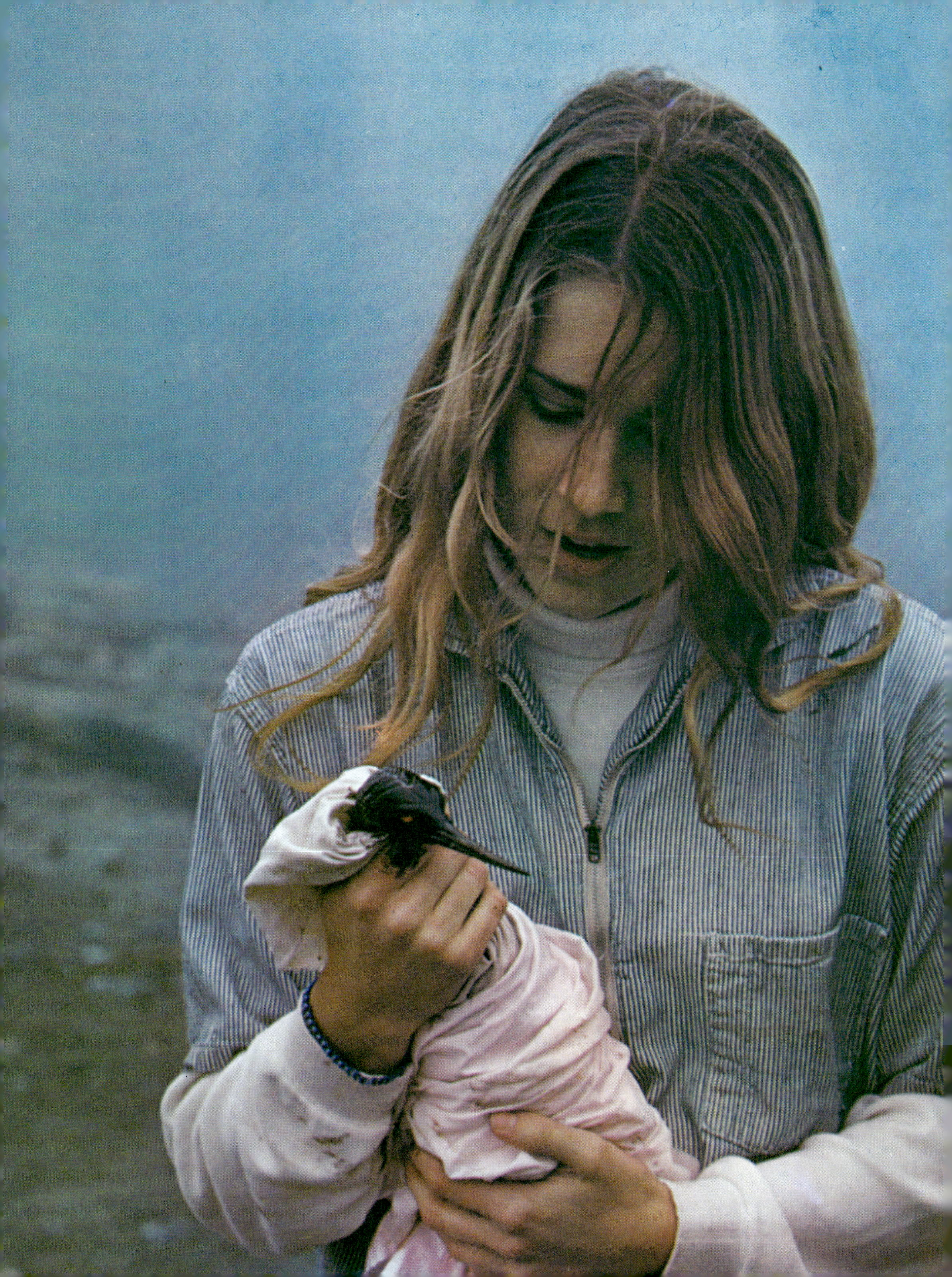

Spectacular Return

Each time a species of marine mammals has been really protected, it has made a spectacular comeback. Examples of this encouraging efficiency are the sea otter, right whale, gray whale, and elephant seal. The elephant seals had the narrowest escape.

The hunters went after them with muskets, spears, and clubs, slaughtering them year after year until by the turn of the century there were fewer than 100 northern elephant seals left in the entire world. Other hunters, stopping at various subantarctic islands, massacred the southern elephant seals at a similar rate, reducing their population from well over a million to a few hundred. There were so few northern and southern elephant seals that it was no longer economically worthwhile to hunt them. Hunters could get close to 100 gallons of oil from a large bull and one unusually large male yielded 210 gallons. As soon as hunting in the southern seas stopped for lack of animals, their population began to increase. Most lived on islands with few if any people on them, so they were able to breed unmolested and today their population is sizable. Sealing is carried on under strict supervision with a quota of 6000 bulls of 10½ feet or longer allowed the sealers.

The northern elephant seals, which once ranged 1000 miles south from San Francisco Bay, were hunted from 1818 until the 1860s. By then there were so few it just was not worth the effort for the few seals a hunter could kill. A small herd of elephant seals remained by 1880. Over the next few years another 300 or so were killed completing the job, everyone thought, of exterminating the northern elephant seal. The seals were no help toward the survival of their the decision, an armed garrison was established on the beach of Guadalupe Island. Today between 10,000 and 15,000 northern elephant seals live on Guadalupe, San Benito, and San Geronimo islands off Mexico and in the Santa Barbara Islands and Los Coronados off southern California. As their comeback gained momentum problems of overpopulation ensued. Too many cows and bulls crowded the narrow beaches leaving little room for the helpless pups which became easily trampled. Eventually though there was movement to other islands and a balance of sorts was restored. The existing populations are all descendants of the Guadalupe Island 100.

A herd of sea elephants lie at water's edge on a rocky beach. Behind is a sheer cliff that serves as protection.

Care of young. Sea elephant cow and its pup mix in with other cows and pups on a beach.

race. Their naturally lethargic behavior and phlegmatic disposition made them easy targets. All a hunter had to do was approach the big seals from the seaward side and drive them inland, beating, shooting, or stabbing some of them and panicking the others into trampling many pups and cows to death.

More than 20 years later a group of fewer than 100 elephant seals was found living on the island of Guadalupe in the Mexican waters of the Gulf of California. They had survived because the beach was isolated and backed with a 3000 foot cliff. Of these few animals 14 were killed and shipped to be stuffed and displayed as one of the world's rarest species.

The Mexican government ordered the killing stopped in 1911, and by 1922, with the revolutionary government well established, seals were fully protected by law. To enforce

Success of the Fur Seals

Three times in less than 200 years the northern fur seals have been threatened with extermination. And each time, thanks to the foresight of several governments, the fur seals have come back, until today on the Pribilof Islands they form the largest aggregations of mammals in the world.

When in 1786 the Russian explorer and fur trader Gerasim Pribilof discovered the islands in the Bering Sea that now carry his name, they were aswarm with northern fur seals. But with the discovery the slaughter of seals on the rocky beaches began. By 1806 the species was so reduced in numbers that the Russians gave up the killing. The fur seal population responded by rapidly increasing, and the massacre resumed until 1834. Again the stocks of fur seals became so low that the Russians quit the hunt. Again the population built up, and the harvest of fur seal pelts resumed. But the Russians apparently had learned a lesson from the moratoriums on killing they had been forced into. They established rules for the taking of the animals to assure a continuing supply of the fur seals with their rich pelts. According to the regulations set down by the Russians in the 1840s, only bulls on the beach could be taken by the sealers. The policy was so sound that it is still in force today.

In 1867, however, Russia sold Alaska to the United States for $7.2 million. In less than 30 years, a commercial sealing company took more than two million skins and paid the U.S. government enough in taxes and royalties to cover the cost of Alaska's purchase. Planning ahead, the company had set a quota of 100,000 seals a year when it took over the sealing operations. They retained the "bulls only" regulation, and they reaped a handsome harvest until sealing schooners set out to circumvent the company's rules by taking seals at sea. The pelagic sealers poached as many animals as they wanted, and since they were unable to determine sex when the animals were in the water, they frequently took more cows than bulls. Many females were pregnant. The toll was tremendous as more and more pelagic sealers took to their ships. Up to 80 percent of the seals they killed at sea were lost when they sank. Again the species was threatened.

To halt this senseless killing, Russia, United States, Japan, and Great Britain signed an agreement. Canada joined later. The treaty permits taking fur seals on the beaches of their breeding islands, which include the U.S.-owned Pribilofs, the Russian-owned Kurile Island chain, and the Commander and Roben islands. Japan, Britain, and Canada each receive a share of the killed seals as compensation for not taking them at sea. Only bachelor bulls, easy to separate from the rest of the seal population, may be taken.

Today the estimated population of northern fur seals is over two million. With continued protection, the fur seals are here to stay.

There are seven other species of fur seals, mostly in the southern hemisphere. An eighth species from the Juan Fernandez Islands off the Chilian coast was until recently thought to be exterminated. In 1973 a small, healthy colony was discovered and is fully protected and promising another recovery. The others were severely decimated, hovering near extinction in several cases before beginning a long, slow recovery. The Guadalupe fur seal found in the Gulf of California sharing an island with elephant seals, was thought extinct for a number of years. Then a small population was discovered, and through cooperation between the U.S. and Mexico, the species is beginning to thrive.

Fur seals line up in water off the beach on State Island in the Pribilofs.

Gray whales, among the slowest swimming cetaceans, were an easy mark for harpooners and gunners seeking them for their oil. The whales cruise at a speed of only four miles an hour—about the speed of a man walking. Their top speed of eight to ten knots is not much more than that of a human runner.

The result of the gray whale slaughter on both sides of the Pacific was spectacular. From a population of about 25,000 the gray whales were reduced to fewer than 200 on the American side of the ocean and probably not much more than that on the Asian side. Shortly after the International Whaling Commission was first convened in 1946, gray whales were placed under protection by the IWC. The only exceptions were that 50 gray whales a year are permitted the native population of Kamchatka peninsula in

*The **slow-swimming gray whale** (left) was an easy mark for harpooners and their numbers were greatly reduced before enforced protection.*

Protection Pays Off

The 10-foot-high blow of gray whales made them easy to see for the Indians of the west coast of North America as they hunted them over the past centuries. The whales' coastal habits put them within easy range of the Indians and their small boats. But the Indians, who depended on the gray whales for part of their existence, did not decimate the grays. They took only what they needed and used all that they took, feeding on fat and flesh and extracting oil from the remains. Just before gold was discovered in California in the late 1840s, intensive hunting for the grays began off that coast. Whaling ships sailing from New Bedford had to round Cape Horn to get into the eastern Pacific where the grays dwell. Half a century later, a serious fishery for gray whales began to develop off the Korean coast in the western Pacific.

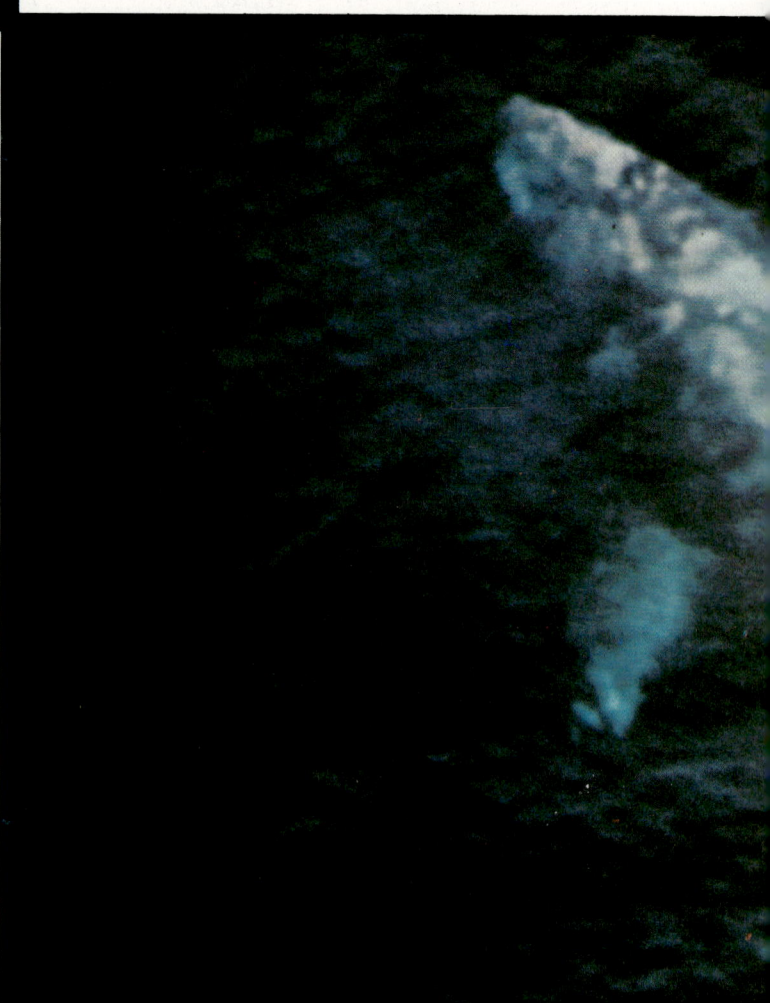

northern Siberia, and that 100 are allowed to the U.S. government to study the gray whale population. The IWC ban has been enforced and their numbers are increasing as dramatically as they once declined.

Protection from man, however, is not the only reason for the continuing comeback of the gray whale. These plankton-eaters, which grow to a maximum length of 45 feet, breed very readily. Cows can conceive almost immediately after giving birth; this means they can calve once a year, giving birth to a 15-foot baby weighing up to 1500 pounds.

Now, as the gray whale population in the eastern Pacific edges up to the 10,000 mark, it is clear that when protective laws are seriously enforced, endangered species of marine mammals can be saved.

Annual migration of gray whales (below) to and from breeding grounds is viewed by thousands of watchers. Here are a pair along the California coast.

A **barnacle-encrusted gray whale** (above) surfaces for air as it migrates along the California coast toward its breeding grounds in Baja California.

The Way of Right Whales

Off the Argentine coast, where the southern right whales gather to frolic, to mate, and to tend their calves, can be seen some of the most touching scenes of motherhood among wild creatures. A calf, breaching and blowing around its mother, bumps into her repeatedly. The 20-foot-long baby bounces on patient and stolid mama one time too many. The cow, perhaps 45 feet long, rolls onto her back, and when the calf breaches again, she gathers it in with a flipper to her chest and holds it for a while, until the calf has lost some of its exuberance. Then she gently releases her hug, and the calf quietly snuggles up to its mother's side. Scenes like this were commonplace at sea when right whales were abundant all over the world, but that was a long time ago and no longer the case.

Today all three species of right whales are scarce. The Argentine coast whales are about the same as those found in many parts of the Northern Hemisphere, especially around the Bay of Biscay where the Basques of eight centuries ago used to go whaling. Biscay right whales grow to about 50 feet in length.

The other large-sized species of right whale is the bowhead, or Greenland right whale, which lives in arctic waters east and west of Greenland. These are the most grotesque of all the whales with their jaws arched up almost into a semicircle. They grow to a length of 60 feet, and like the Biscay right whales, they are slow swimmers. The baleen and blubber made them most attractive to the early American whalers who decimated the sluggish and trusting creatures. By the beginning of the twentieth century the situation was so grim for the bowhead that an international agreement was reached to halt killing of Greenland right whales. The Biscay species had to wait until 1929 before an